Integration and self regulation of quality management in Dutch agri-food supply chains

Project part-financed by the European Union (European Regional Development Fund) within the InterregIIIC Programme, project PromSTAP.

Integration and self regulation of quality management in Dutch agri-food supply chains

A cross-chain analysis of the poultry meat, the fruit and vegetable and the flower and potted plant chains

Wijnand van Plaggenhoef

International chains and networks series – Volume 4

Wageningen Academic
P u b l i s h e r s

ISBN 978-90-8686-055-5
ISSN 1874-7663

The individual contributions in this publication
and any liabilities arising from them remain the
responsibility of the authors.

First published, 2007

Wageningen Academic Publishers
The Netherlands, 2007

International chains and networks series

Agri-food chains and networks are swiftly moving toward globally interconnected systems with a large variety of complex relationships. This is changing the way food is brought to the market. Currently, even fresh produce can be shipped from halfway around the world at competitive prices. Unfortunately, accompanying diseases and pollution can spread equally rapidly. This requires constant monitoring and immediate responsiveness. In recent years tracking and tracing has therefore become vital in international agri-food chains and networks. This means that integrated production, logistics, information- and innovation systems are needed. To achieve these integrated global supply chains, strategic and cultural alignment, trust and compliance to national and international regulations have become key issues. In this series the following questions are answered: How should chains and networks be designed to effectively respond to the fast globalization of the business environment? And more specificly, How should firms in fast changing transition economies (such as Eastern European and developing countries) be integrated into international food chains and networks?

About the editor

Onno Omta is chaired professor in Business Administration at Wageningen University and Research Centre, the Netherlands. He received an MSc in Biochemistry and a PhD in innovation management, both from the University of Groningen. He is the Editor-in-Chief of The Journal on Chain and Network Science, and he has published numerous articles in leading scientific journals in the field of chains and networks and innovation. He has worked as a consultant and researcher for a large variety of (multinational) technology-based prospector companies within the agri-food industry (e.g. Unilever, VION, Bonduelle, Campina, Friesland Foods, FloraHolland) and in other industries (e.g. SKF, Airbus, Erickson, Exxon, Hilti and Philips).

Dedicated to my parents

Table of contents

Acknowledgements

Conducting a PhD study is a wonderful time and an opportunity for all who have the chance to do it. However, this thesis would not have been completed without the help of many people. Words are insufficient to thank them for all the constructive criticism and advice they have offered me during the four years.

First of all, I acknowledge the Dutch Ministry of Agriculture, Nature and Food Quality (LNV, Department of Industry and Trade) and the Food and Consumer Product Safety Authority (VWA) for partly funding this study. I vividly remember the meetings my supervisors and I had with Roel Bol (Director of the Department Industry and Trade of LNV), Wim de Wit (former Director of the VWA) and Marcel Mengelers (expert of the Department Risk Assessment of the VWA) in The Hague each half year. Thank you for all your good suggestions, great interest and support!

Many thanks also go to Jos Ramekers, chairman of the Product Boards of Livestock, Meat and Eggs and Jaap van der Veen, former chairman of the Product Board for Horticulture, for supporting the study. Without their kind help, this study would not have been possible. I also want to thank all those people who were willing to participate in the interviews or filled out the questionnaires. I hope the results will help you in the management of your daily work.

I am greatly indebted to my supervisors Onno Omta, Paul van Beek and Jacques Trienekens. Onno thank you for stimulating me to go still one step deeper every time. You taught me the academic skills necessary for conducting research and your comments gave this thesis the high academic level. Paul you always stressed the practical aspects of the study by asking questions such as: 'What do you exactly mean with this sentence?' Thanks also for the constant interest in my personal and professional well-being. Jacques, thank you for answering all my questions during the day-to-day support and for the introduction in your network of supply chain experts. It was always possible to enter your office for some consult. Thank you all!

Furthermore, my gratitude also goes to Ron Kemp and Ivo van der Lans for answering my methodological questions, especially with regard to Lisrel. Sometimes, you must have become slightly mad of all my questions. Many thanks also go to William Banfield for correcting my English.

I also want to thank my colleagues in the Management Studies Group of Wageningen University. It was a pleasure to work with you all; you gave me new ideas about the study and always offered support for solving problems. In particular, thanks to Maarten Batterink for the nice time we had in room 5055 in the Leeuwenborch and for all the deep discussions we had.

Tenslotte wil ik ook mijn naaste familie bedanken voor alle ondersteuning die zij mij gedurende het onderzoek hebben gegeven. In het bijzonder wil ik mijn ouders bedanken voor alles wat zij voor mij gedaan hebben. Bedankt voor een opvoeding waarin eenvoud en no-nonsense centraal stonden. De lege plek die pa heeft achtergelaten is na bijna tien jaar nog steeds voelbaar. Hij zou hier graag bij aanwezig zijn geweest. Ma, ook bedankt voor het aanhoren van al '*dat gemauw*' en dat je altijd weer de kans zag om mij op te beuren als dat nodig was. Als er ooit een prijs komt voor de beste ouder van Nederland, ben jij in ieder geval genomineerd!

Wageningen/Nijkerkerveen, August 2007

Chapter 1. Introduction

The first chapter of this book discusses the rationale for conducting a study about integration and self regulation of quality management in agri-food supply chains. Section 1.1 presents a general introduction and describes the current developments with regard to quality management and self regulation in Dutch agri-food supply chains. Next, the research questions are described in Section 1.2. The theoretical and managerial contributions of the study are provided in Section 1.3. The selection of the chains in which the study is conducted is motivated in Section 1.4. Section 1.5 describes the advantages of a 'mixed methodology' (combining quantitative and qualitative approaches) in the study. The chapter ends with an outline of the book in Section 1.6.

1.1 General introduction

The present study deals with the integration of quality management in agri-food supply chains. Although research interest in supply chain management is clearly growing, only few studies have been focused on quality management practices in a supply chain management perspective (Robinson and Malhotra, 2005).

During the last decade, concerns about quality and safety in agri-food supply chains have been raised among consumers. Several sector-wide crises, such as the BSE crisis, the dioxin crisis, classical swine fever, foot-and-mouth disease and Aviaire Influenza have fuelled these concerns and indeed, when quality assurance fails, the adverse consequences can be large. For example, it is estimated that in the United States alone, contaminated food causes up to 76 million illnesses, 325,000 hospitalisations and 5000 deaths each year (Smith-DeWaal, 2003). Through media coverage consumers in industrialised countries have become more aware of the potential hazards (Jouve, 1998; Opara and Mazaud, 2001; Unnevehr and Roberts, 2002). Mass media and specialised publications propagate transparency about agri-food firms' quality assurance to the public (Frombrun and Shanley, 1990).

The crises have also increased consumer awareness of (other) side effects of bio-industrial production. As a result, the concerns of consumers may not only be limited to safety and quality issues, but also important ethical concerns are raised, for example, concerning the destruction of animals associated with the BSE crises (Van Kleef *et al.*, 2006). Due to all this attention, consumers have become more critical regarding the food products they buy. Nowadays, consumers demand more information about the origin and the safety of their food, including the means of production, hygiene, genetic modification, application of pesticides and other environmental issues.

Chain-wide integration of quality management systems is regarded as the best strategy to deal with these complex quality demands because no individual firm is able to handle quality on its own (Omta *et al.*, 2002). This vision is strengthened by a study of Cap Gemini and Ernst & Young (2002) in which a vast majority of the managers of European food processors (86%) and retailers (87%) indicates that food quality is basically a task of the agri-food supply chain as a whole. If food problems arise and recalls are necessary all parties in the chain will be affected, therefore, all supply chain partners should take their responsibility to assure the quality of food (Grievink *et al.*, 2003). As a result quality has become an integral element of most farmers', wholesalers' and retailers' business strategy (Antle, 1999).

Firms increasingly respond to their tasks of quality assurance by adapting (private) quality management labels in which firms ask their suppliers to comply with certain regulations (Freriks, 2006). Big retailers in particular have developed initiatives to commit their suppliers to strict food safety regulations. These quality management systems rely on documentation of production processes, combined with third party auditing and certification placing strong requirements on gathering, storing, processing and transfer of quality information between the firms in the chain (Jahn *et al.*, 2004). These private initiatives to regulate and to improve food safety and quality could be regarded as forms of self regulation, also known in literature as 'self enforcing', 'self governance' and 'self organising' (King and Lenox, 2005; Havinga, 2006).

(Inter) national governmental agencies have also reacted to the above mentioned crises by setting up regulations for quality and safety of agri-food products. For example, the European Union has issued the General Food Law in which the primary quality responsibility of firms in agri-food supply chains is emphasised. In agri-food supply chains, many firms go beyond compliance with legal regulations, because they have to meet the expectations of their buyers and avoid reputational disasters (Bondt *et al.*, 2006; Freriks, 2006; Havinga, 2006). However, concerns have been raised about the burdens (especially administrative) being placed on the firms, because at the moment for many issues firms have to comply with (inter) national quality regulations and with private quality regulations. For example, for firms in fruit and vegetable chains the monitoring of pesticides residuals on products is an important issue in the Pesticide Law, but also in the private quality system Eurep-GAP.

In order to reduce the compliance burdens, the Dutch Ministry of Agriculture, Nature and Food Quality strives for a new inspection policy, called '*control-on-control*', where the private sector is assigned more responsibility for compliance with statutory quality and safety regulations. '*Control-on-control*' fits within the changing policy of the Dutch government in which the role of the government shifts from one-sided practices in which government was solely responsible for strategic planning ('*command-and-control*' approach) towards improving governance and creating transparency in policy processes. One outcome of this process was a rearrangement of the balance between public and private responsibilities, also with regard to quality assurance. This has led to Public Private Partnerships, as well as to a more performance-oriented government. At the same time, agri-food firms changed their strategic focus from cost to adding value for the buyer. These developments have far-reaching consequences for the way quality assurance in agri-food supply chain processes is structured and managed, as well as for the roles and responsibilities of the different actors involved. 'Control-on-control' is an example of this new policy in which firms receive the objectives and the conditions to fulfil a certain policy. In practice this means that firms that perform well on quality management will receive lower inspection frequencies than bad performing firms (LNV, 2004b). In this new situation the Ministry operates at a greater distance but retains the ultimate responsibility, because even if the vast majority of firms do the right thing, there is always the chance that a firm will produce serious harm. The Dutch Ministry expects that 'control-on-control' will result in a more efficient and effective assurance of the firms' interests (De Bakker *et al.*, 2007).

1.2 Research questions

Even though much attention has been paid to supply chain management and quality management during the last decades, the interlinking between these is often limited and tangential in nature. The strict quality requirements set by retailers, which are often more relevant to the quality management systems of firms than those set by the government, are one of the incentives for the emergence of tightly coupled agri-food supply chains in which quality information is transferred (Grabosky and Gunningham, 1998). Moreover, a number of other factors have an impact on integration of quality management systems in chains such as the prevention of media attention or the quality strategy of a firm. The essence of closely integrated chains is to create collaboration in which partners share information, work together to solve problems, jointly plan for the future and make their success interdependent (Krause and Ellram, 1997; Spekman *et al.*, 1998; Shin *et al.*, 2000). Due to the higher transparency in such chains, they effectively achieve a common interest to comply with quality regulations (commitment) and have the means to sanction each other in case of non-compliance (enforcement), which are the two most important dimensions of compliance behaviour in self regulating systems (Balk-Theuws *et al.*, 2004).

When the interests of the buyers and suppliers in the chain coincide, awareness of and knowledge of how to do things right will foster commitment. This kind of commitment helps to develop acceptance and responsibility and strengthen compliance with quality requirements. Havinga (2006) adds to this that the best conditions for private regulations are when firms act responsibly and are willing to comply with reasonable rules. On the contrary, with regard to sanction instruments, Grabosky and Gunnigham (1998) have argued that they should be used with caution, given the costs entailed in implementing them, the uncertainty involved in mobilising them successfully and the risk that if their use is perceived to be unjust or unreasonable, they can trigger a backlash which only works detrimentally. Therefore, enforcement, including sanctions, should be used as a last resort, only necessary for the ones for which neither information, incentives, nor compensation were sufficient to comply with the regulations (Grabosky and Gunningham, 1998). These remarks are in line with the conclusions of Morgan and Hunt (1994) that commitment and not enforcement is the best way to condition the behaviour of firms.

Regarding the important role of integration and self regulation of quality management, understanding how firms integrate their quality management systems with their buyers and suppliers is perhaps among the most essential questions for agri-food firms. Integration and self regulation would aim to fulfil the goals of providing high buyer value with an appropriate use of resources i.e. to enhance performance. However, fostering and maintaining a good relationship between partners is a daunting task, so the understanding of the factors that determine the successful integration of self regulated quality management systems in agri-food supply chains is really important (Fuentes-Fuentes *et al.*, 2004). Self regulation is one of the main topics in public private partnership discussions nowadays. In order to address the problems described above, four research questions can be formulated:

1. Which internal and external factors have an impact on the integration of quality management systems in agri-food supply chains?
2. How does integration of quality management systems affect self regulation in agri-food supply chains?
3. How do integrated quality management systems affect performance in agri-food supply chains?
4. What is the best way to create self regulated quality management systems in agri-food supply chains?

A number of theories deal with collaboration in supply chains, which are useful in this study. Supply Chain Management describes how business transactions are conducted in supply chains (Lambert *et al.*, 1996; Lambert and Cooper, 2000). For the management of quality, the study focuses on Total Quality Management and creates an interlinking with Supply Chain Management. For choosing an appropriate governance form in the supply chain, Transaction Cost Theory (Rindfleisch and Heide, 1997; Bijman, 2002) and Contingency Theory (Lawrence and Lorsch, 1967) provide important theoretical insights.

1.3 Theoretical and managerial contributions

The quality management perspective in supply chains has received only limited research attention up to now, even though that perspective is surely needed to deliver value to the buyers in, often globally scattered, agri-food supply chains. In a recent study, Robinson and Malhotra (2005) have stressed the necessity to translate quality practices from a traditional firm based approach to an inter-organisational supply chain orientation involving both buyers and suppliers of the focal firm[1]. The effect of integration and self regulation of quality management on performance is also interesting, because empirical studies have produced mixed results so far (Samson and Terziovski, 1999). Another contribution is that this study is one among

---

[1] Focal firms are the firms where the present study has been conducted; i.e. managers that participated in the interviews or filled out the survey.

the few quantitative studies that test the validity of the key prediction of buyer supplier relationship management given by Morgan and Hunt (1994), namely that commitment more than enforcement, will lead to more successful buyer-supplier relationships with regard to quality management.

The present study will provide managers with a practical insight into which factors contribute to successful integration of quality management systems with their buyers and suppliers. In addition, it is shown to what extent the integration of quality management and self regulation contribute to a firm's performance. Integration of quality management is interesting for policy makers too, because it reveals what is the best way to organise self regulation in integrated chains. This perspective fits with the strategy of government in which government and industry are not each others' opponents, but aim to achieve joint goals (LNV, 2004b). Answering the last research question will provide policy makers and managers with practical examples on how integrated and self regulated quality management systems could be effectively designed in agri-food supply chains.

1.4 Study domain

This study is carried out in the poultry meat chain, the fruit and vegetable chain and the flower and potted plant chain for the following reasons (Deneux *et al.*, 2005; Berkhout and Van Bruchem, 2006):
1. They are valid representations of the agri-food sector, because of the high diversity of marketing channels and products. The poultry meat chain is the most strongly integrated chain with a limited number of processors which sell their products mainly through big retailers. The same holds to a less extent for the fruit and vegetable chain. However, in the fruit and vegetable chain no serious food safety crisis has occurred in the past. Contrary to the other two chains, the flower and potted plant chain is somewhat different with regard to marketing channels, because flowers and potted plants are mainly sold by small outlets and large retailers. Another difference is that, compared with the other two chains, in the flower and potted plant chain food safety management, as a part of quality management, does not play a role.
2. They are all of big interest for the Dutch economy. The flower and potted plant chains export products with a value of more than 5 billion Euros each year. The fruit and vegetables chains also exports 4.4 billion Euros per year, while the export of poultry meat was almost 1 billion Euros per year.
3. They are among the chains that pay the most attention to quality management. Quality issues in the poultry meat and fruit and vegetable chain mainly deal with food safety, while in the flowers and potted plants chains issues regarding environmental management, labour and health circumstances are especially dealt with. In all three chains many private quality management systems have been developed and implemented, of which more than 90% of the firms participate.

A special feature of this study is that it collects data from both the supplier and the buyer side of the focal firm and includes two successive firms in each chain (primary producers and traders and/or processors). This approach ensures the suitable implementation of the Supply Chain Management approach, which, in its broadest sense, spans the entire chain from initial source (supplier's supplier, etc.) to ultimate consumer (buyer's buyer, etc.). Most previous studies were limited to data collection in one firm or took only the buyer or the supplier side of the focal firm into account.

1.5 Methodology: three phases

The accuracy of studies in the field of management and organisations can be improved by collecting different kinds of data about the same phenomenon. A 'mixed methodology' combines qualitative and quantitative research methods and is also known as *triangulation* (Jick, 1979; Verschuren and Doorewaard, 1999; Tashakkori and Teddlie, 2003b, a; Teddlie and Tashakkori, 2003). The assumption of a 'mixed methodology' is that weaknesses of one method can be compensated by the strengths of other methods. For example, a survey overcomes the limited generalisability or external validation of case studies, while case studies provide insight in the way of working of firms in their single, natural setting, things that are hard to include in surveys. In this way, a 'mixed methodology' ensures that the variances and observations represent the phenomenon of interest and not the method used, offering a greater potential for consistent theory building (Wacker, 1998).

The *first* phase of this research started with the identification, description and ranking of external drivers acting on agri-food supply chains. Only few studies have been done to explore the impact of the business environment on quality management (Fuentes-Fuentes *et al.*, 2004). Important questions in these interviews are: *Which drivers are acting in the business environment of agri-food supply chains? How do they affect quality management? How important are these drivers and are there any differences in the impact of these drivers in different agri-food supply chains?* In order to answer these questions 47 interviews with experts from business and academia were held and can, as such, be regarded as '*explorative interviews*'. During the interviews a conjoint analysis was used, a quantitative method to come up with a ranking of the importance of the different drivers from the business environment.

In the *second* phase a large scale survey was conducted among primary producers, processors and traders in the three chains, on which almost 600 firms responded. The primary goal of the survey was testing hypotheses. The survey reveals how, and to what extent, factors influence the integration of quality management, how integration of quality management systems influences compliance behaviour and how integration and self regulation of quality management impact performance of firms in agri-food supply chains. Further, it was possible to test the generalisability of the findings across the chains.

In the *third* phase, the findings from the quantitative part of the research were verified using 14 in-depth interviews with primary producers, processors and/or traders in the three chains and experts from certification organisations and governmental agencies. These interviews can be regarded as '*confirmative interviews*'. The objective of this phase is to gain feedback on the results and to get more practical insights in how the relationships found in phase two exactly work in practice. The statistical findings from phase two are associated with statements in the interviews. Another objective of this phase is to come up with examples of 'best practices' about the way quality management and self regulation systems could be organised and provide recommendations for managers and policy makers based on these 'best practices'.

1.6 Outline of the book

This book consists of nine chapters and is divided into four parts, as visualised in Figure 1.1. The *first* part introduces the importance of this study and presents the theoretical rationale. Chapter one introduces the present study. Chapter two provides facts and figures about the three agri-food chains; such as the structure, the number of firms, the most important products and the major export countries for each chain in which the study was carried out. The most important general and chain specific quality regulations and quality systems are also described in this chapter. Chapter three reviews the relevant theories that deal with integration and self regulation of quality management with buyers and suppliers. Important theoretical contributions are extracted from Supply Chain Management, Total Quality Management, Transaction Cost Theory and Contingency Theory. The chapter is complemented with a discussion on the concept of self regulation and ends with a research model. Chapter four deepens the elements of the research model and based on the theories discussed in chapter three, it presents the hypotheses.

The *second* part consists of Chapter five and describes the research design. Firstly, the design of the conjoint analysis is outlined. Secondly, the design of the survey is presented including data collection and measurement instruments, as well as the methods of data analyses. Thirdly, the design of the in-depth interviews is described including the topics covered and the selection of respondents.

The *third* part shows the results of the conjoint analysis, the survey and the in-depth interviews. In chapter six, the conjoint analysis ranks the drivers from the business environment that experts consider to be important to stimulate integration of their quality management with those of their buyers and suppliers. Chapter seven describes the response to the survey and tests the validity, reliability and generalisability of the constructs derived from the research model. After that the estimated models to test the hypotheses are presented for the buyer and supplier side of the focal firm. Furthermore, a number of control variables, such as size, number of quality management systems and presence of a quality manager are analysed. The chapter ends with findings from a number of models which test the hypotheses in the sub-groups and the effect of group specific control variables. In chapter eight the results of the in-depth

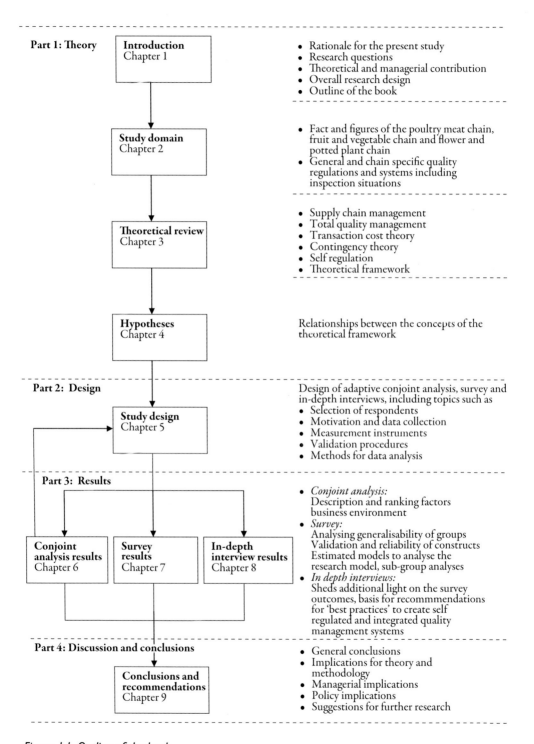

Figure 1.1. Outline of the book.

interviews are described, which shed light on the relationships found in the estimated models and provides a basis for recommendations for managers and policy makers on how to develop integrated and self regulated quality management systems in agri-food supply chains.

The *fourth* part consists of the discussion and conclusions. Chapter nine discusses the implications of the outcomes of the estimated models. The information obtained in the conjoint analysis and the in-depth interviews is included in this interpretation. In addition, recommendations for establishing self regulated and integrated quality management systems are formulated. Finally, the theoretical contributions and suggestions for further research are provided based on the conclusions and limitations of the present study.

Chapter 2. Study domain

This chapter describes the study domain. It starts with a description of the main characteristics of the Dutch agri-food complex in Section 2.1. Furthermore, this chapter focuses more in-depth on the characteristics of each of the three chains included in this study, such as the number of firms, most important products and export markets in Section 2.2 to 2.4. After that, the main European legislation on quality and safety is discussed in Section 2.5. A number of important quality management systems of European retailers are presented in Section 2.6. In Section 2.7 the national quality and safety regulations in The Netherlands are discussed. In Section 2.8 to 2.10 the public and private quality and safety regulations, including the inspection situations for firms are discussed. The chapter ends with some concluding remarks in Section 2.11.

2.1 The Dutch agri-food sector

The total Dutch agri-food sector[2] consists of two parts, representing the economic contribution of processing, delivering and distribution based on *domestic* and on *foreign* agricultural raw materials. The total agri-food complex realises an added value of 9.3 % of the total Dutch economy and 10.1% of the total Dutch employment (see Table 2.1).

The employment of the total Dutch agri-food sector has declined from 751,000 to 651,000 employees during the period 2001 to 2004, while total Dutch employment has increased. Thus, it can be concluded that the importance of the agri-food sector has declined for the total Dutch economy during these years. The added value of the part based on domestic agricultural raw materials had decreased by 1.1 billion Euros during the period 2001 to 2004, while the share in the total Dutch employment had decreased from 6.6% to 5.9% in the same period. In particular, the number of small farms has declined rapidly over the past few years. An important reason for this is the increasing pressure on economic margins.

Grassland-based livestock contributes most to the added value and the employment of the agri-food sector based on domestic raw materials (see Figure 2.1). The share of the arable and intensive livestock complex has declined as a result of the low prices in 2004 for many agricultural products. In the poultry sector in particular the prices have been low. The individual sectors for greenhouse gardening, arable farming and intensive livestock each generate about one fifth of the total added value. Due to the capital intensive character of the greenhouse gardening, its value added per employee is above the average of the total agri-food sector. In spite of this, the sectors open-ground gardening and grassland-based livestock are relatively labour intensive.

---

[2] Information on the economic characteristics of the Dutch food and beverage industry, which is presented in this section, has been mainly obtained from the Dutch Agricultural Research Institute (www.lei.wur.nl).

Table 2.1. Characteristics of the Dutch agri-food sector in 2001 and 2004 in billion Euros (Berkhout and Van Bruchem, 2006).

	Added value (billion Euros)		Employment (billion Euros)	
	2001	**2004**	**2001**	**2004**
Total agri-food sector*	40.5	40.4	714	651
Share in national	9.4%	9.3%	11.1%	10.1%
Gardening, forestry agricultural services	3.6	3.8	71	64
Processing, delivering, distribution of foreign raw materials	14.8	15.6	220	205
Agricultural complex based on domestic raw materials				
Share in national	5.1%	4.8%	6.6%	5.9%
Primary production	7.9	6.9	186	176
Processing	3.3	3.4	53	45
Delivery industry	7.9	8.4	130	122
Distribution	3.0	2.3	54	40

*Including gardening, forestry, agricultural services, cacao beverages and tobacco.

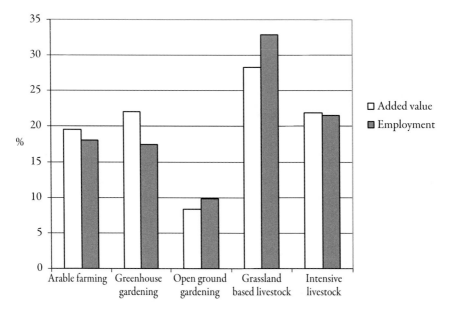

Figure 2.1. Shares (%) of value and employment in the different sectors of the Dutch agri-food sector in 2004 (Van Leeuwen, 2006).

The Dutch agri-food sector also processes many foreign raw materials, such as materials for feed processing, grains for human consumption, oil seeds, cacao, coffee and tea. The added value of this part has increased during the period 2001 to 2004 to 15.6 billion Euros, but employment has decreased with 15.000 full time equivalents (Table 2.1.). Regarding these figures, this part of the Dutch agri-food sector retains its share in the Dutch economy. The employment of the agri-food sector based on domestic agricultural raw materials counted for approximately 60% of the total employment of the total Dutch agri-food sector, whereas the agri-food sector based on foreign raw materials counted for the remaining 40%. However, the food processing industry based on foreign based raw materials has generated 70,000 jobs more than the food processing industry based on domestic raw materials.

There is a strong dependency of the Dutch agricultural complex on exports. In total the Dutch agri-food sector exported for almost 51 billion Euros, which means a share of 18% in the total Dutch export in 2005. The importance of flowers and potted plants in particular has increased. Flowers and potted plants are the most important export products and belong to the most competitive products. Most competitive products are defined as having a more than average export share on the world market. According to this definition, approximately half of the 100 most competitive export products belong to the agri-food sector. 82% of agri-food products are exported to EU countries. The most important export country is Germany, followed by the United Kingdom, Belgium and France. Regarding import, the ranking of the most important trading partners is the same. The main import products of The Netherlands are dairy and meat products. The total value of imports was 28.2 billion which represents 11% of the total Dutch import.

Most agri-food products are sold in supermarkets (see Figure 2.2). In 2005, consumers spent 23 billion Euros on agri-food products in supermarkets. During the last few years, supermarkets have broadened their assortments, especially regarding the fresh segment. Only organic and foreign foods specialty stores can retain their market shares. Besides these two product groups a considerable amount of fish is sold by means of traditional speciality stores.

Regarding the market share of supermarkets in The Netherlands, Albert Heijn is the most important supermarket (see Table 2.2). It also has the largest market share for fresh products, because of its large sales of potatoes, fruit and vegetables, fresh fish and convenience foods. For the last category Albert Heijn even realises a market share of 60%. C1000 and PLUS are other supermarkets with large fresh assortments. Also the discounter Lidl has increased its assortment of potatoes, fruit and vegetables during the last years.

Having described the main characteristics of the Dutch agri-food sector above, the next sections will focus more in-depth on the chains that are the subject of the present study, the poultry meat chain, the fruit and vegetable chain and the flower and potted plant chain.

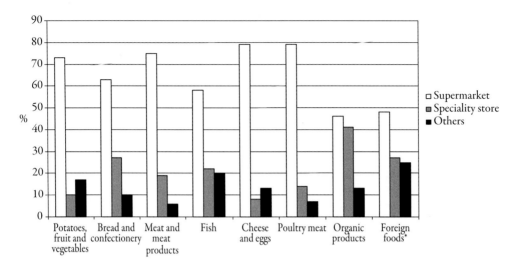

Figure 2.2. Market share (%) of different agri-food products in 2004 (Detailhandel, 2005).
*Data for 2000

Table 2.2. Annual turnover, number of stores and market share of important Dutch supermarkets in 2005 (Berkhout and Van Bruchem, 2006).

Firm and formulas	Turnover (million €/year)	Store number	Market share (%)
Ahold (Albert Heijn)	6,418	674	26.9
Schuitema (C1000)	3,129	462	14.8
Albrecht (Aldi)	1,500	391	9.5
Laurus (Super De Boer, Edah, Konmar)	3,158	700	12.0
Sperwer Group (PLUS, Spar)	1,150	539	6.2

2.2 The poultry meat chain

Within the poultry chain two important chains exist, the poultry meat chain and the egg chain. However, the poultry meat chain is more interesting than the egg chain with regard to quality assurance, because it faces higher safety and quality risks. Therefore, this study focuses on the poultry meat chain of which the main characteristics such as number of firms and traded quantities are shown in Figure 2.3.

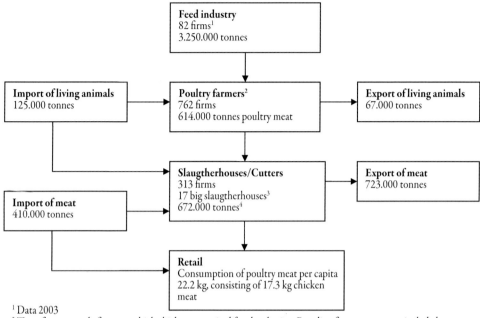

¹ Data 2003
² These firms are only firms on which chickens are raised for slaughering. Breeding firms etc., are not included.
³ Big slaughterhouse: yearly supply > 10.000 ton
⁴ Slaugthered weight
Source: LEI, CBS, PD and PVE

Figure 2.3. The poultry meat chain in The Netherlands in 2005.

The poultry meat chain starts with the feeding industry which produces 3.3 million tonnes of feed; the poultry meat sector uses half of this quantity, and the other half is consumed by the egg sector. A large part of the raw materials for feed is imported. The animal feed sector is highly concentrated. Although 82 firms were involved in feed in 2003, only 9 out of the 82 firms were responsible for 70% of the total feed production. Important feed firms in The Netherlands are Cehave Landbouwbelang voeders (www.cehave.nl), Agrifirm (www.agrifirm. com), Rijnvallei (www.rijnvallei.nl), Arkervaart-Twente (www.arkervaart-twente.nl) and De Heus/Brokking/Koudijs (www.deheusbrokkingkoudijs.nl).

In The Netherlands 3.6 million parent chickens produce 720 million of hatchery eggs per year of which 60% is intended for Dutch poultry meat farmers, 23% is exported as eggs and 18% is exported as living animals. At the poultry meat farms the chickens are raised within six or seven weeks to a weight of approximately 1.5 to 2.5 kg. Usually chickens are housed in closed sheds living on sawdust covered floors with a density of approximately 20 chickens per square metre. Table 2.3 shows that during the last decades poultry farming has become a very concentrated activity; the total number of chickens in The Netherlands has increased while the number of firms has decreased by more than two thirds. In 2005, the average poultry meat

Table 2.3. Number of poultry meat farms from 1975 till 2005 (CBS and LEI, 2006).

Year	Less than 10.000	10.000- 25.000	25.000- 50.000	50.000- 75.000	75.000 and more	Total number of firms	Total number of chickens (in mln)
1975	35.3%	45.7%	15.0%	3.0%	1.0%	2,329	39,250
1985	20.5%	40.6%	27.3%	8.2%	3.4%	1,477	38,383
1995	13.8%	34.5%	31.0%	13.7%	7.0%	1,301	43,827
2000	7.9%	21.2%	34.7%	20.7%	15.5%	1,094	50,937
2001	6.0%	21.4%	35.0%	20.1%	17.5%	1,027	50,127
2002	6.1%	21.1%	33.2%	21.1%	18.5%	1,096	54,660
2003	5.1%	16.5%	35.2%	21.2%	22.0%	777	42,289
2004	5.7%	16.3%	32.7%	20.8%	24.5%	771	44,262
2005	5.9%	15.9%	31.9%	21.8%	24.5%	762	44,496

farmer owned almost 58,000 chickens while in 1975 this was around 17.000. The number of chickens in The Netherlands reached almost 55 million in 2002, but showed a significant drop to 42 million in 2003. This huge decrease was a result of the Aviaire Influenza ('Bird Flu') crises, which especially harmed smaller poultry farmers. Dutch poultry meat farmers realised a production of 614.000 tonnes of poultry meat, including a relative small amount of meat from turkey, laying hens, ducks, geese and guinea fowls.

In 2005 the production of poultry slaughterhouses reached 672.000 (slaughtered) tonnes poultry meat consisting of 617.000 tonnes chicken meat. Table 2.4 depicts the number of slaughterhouses according to their size in 2005 (PVE, 2006). Again, concentration is the key word, the five biggest slaughterhouses in The Netherlands counted for almost half of the production. The main reason for the concentration is cost reduction. Ollinger et al., (2000) have shown that the costs decrease with 5 to 7% when the slaughter capacity doubles. Large slaughterhouses in The Netherlands are for example, Plukon Poultry (www.friki.nl), Storteboom (www.storteboom.nl), Clazing (www.clazing.nl) and Van den Bor Pluimveeslachterijen (www.borpluimvee.com).

The Netherlands is an important exporting country, because it produces roughly twice the domestic consumption of poultry meat. In 2005 there was an export of poultry meat (including cooked and/or canned meat products) of 723,000 tonnes, which consists of 652,000 tonnes of chicken meat. In the first half of the year 2005 exports of chicken, cockerel and broiler meat increased with 3%, but in the second half of 2005, Bird Flu was responsible for a decrease in consumption in South European countries in particular, which negatively affected Dutch

Table 2.4. Number of slaughterhouses in The Netherlands with a yearly supply of more than 10.000 tonnes (PVE, 2006).

Supply categories (in tonnes)	Number of slaughterhouses 2004	2005	Total supply per category (x 1,000 tonnes) 2004	2005
10,000-20,000	4	2	58	31
20,000-30,000	6	6	159	160
30,000-40,000	3	4	99	131
> 40,0000	5	5	284	295

poultry meat exports (PVE, 2006). Figure 2.4 shows the exports of poultry meat from The Netherlands to various countries.

In 2005, countries within the EU counted for 79% of the total export, while this percentage was 67% in 1995. Germany and the United Kingdom are by far the most important export markets, followed by Belgium and France. Because the majority of poultry meat is sold within the EU, it has to comply with comparable quality standards as in The Netherlands. In the past, the share of Germany in the Dutch export has decreased, because Dutch poultry meat

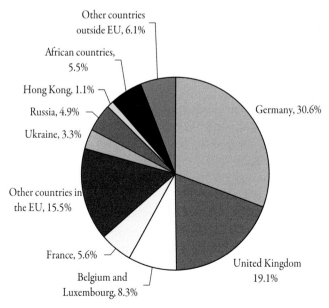

Figure 2.4. Exports of poultry meat from The Netherlands to a number of countries in 2005.

was replaced in favour of imports from countries such as Brazil, Thailand and Poland and due to a higher own production on the German market (WUR Projectgroep Veerkrachtige Pluimveevleesproductie, 2004). However, in absolute quantities the export to Germany has increased. To countries outside the EU export mainly consists of the legs and fowls, whereas filets are preferred in Western Europe. Notwithstanding the status of The Netherlands as an exporter of poultry meat, The Netherlands has imported large quantities of poultry meat as well. The overall imports of poultry meat including cooked and/or canned products reached 410,000 tonnes in 2005. Import of unprocessed chicken meat (excluding cooked and/or canned) represents a total of 276,000 tonnes in 2005. Brazil and Thailand are countries that export a lot of frozen poultry meat to The Netherlands. The United Kingdom also exports a large quantity of poultry meat to The Netherlands, but poultry meat products from this country are mainly used as raw materials for the snack industry (Van Horne *et al.*, 2006).

The popularity of poultry meat has steadily grown as a result of the trends of increasing demands for convenience food and low fat content food. Poultry meat, and in particular chicken filets, benefited from the actions that were organised by supermarkets for this type of meat in the past. In addition, the low price of poultry meat compared with other meat products has also contributed to the increased consumption of poultry meat to 22.2 kg per capita in 2005. Consumption in that year was still 300 grams lower than in 2002, which was a top year for poultry meat consumption (PVE, 2006). Consumption of chicken meat also increased in 2005, to a height of 17.5 kg per capita. In The Netherlands, 84% of the poultry meat is sold in supermarkets. Supermarkets mainly sell pre-packed poultry meat in order to prevent cross contamination of Salmonella and Campylobacter.

2.3 The fruit and vegetable chain

A representation of the fruit and vegetable chain is depicted in Figure 2.5. The fruit and vegetable chain starts with reproduction. Cultivation of breeding material and reproduction of plants, developing new plant species, new upgrading techniques and product innovation including gen technology are important for reproduction farms. At this stage, the basic materials are bred for the growers. Important firms in this stage are Syngenta Seeds (www.syngenta.com), Nunhems (www.nunhems.com), Rijk Zwaan (rijkzwaan.nl), Enza zaden (www.enzazaden.nl), and De Ruiter Seeds (www.deruiterseeds.nl). The last three firms carry out research in the Biopartner alliance (www.biopartner.nl).

The most important vegetables that are grown under glass are tomatoes, peppers and cucumbers. Vegetables grown under glass realise 1.2 billion Euros of total production value (2.2 billion Euros) of fruit and vegetables (including mushroom and onions) in The Netherlands (Frugi Venta, 2006). In 1971 tomato was the most important vegetable grown under glass with 3,000 ha, however, the importance of tomato decreased to 1,380 ha in 2005 (Berkhout and Van Bruchem, 2006). Many growers of fruits and vegetables shifted in the past decades to the growing of peppers; in 1971 only 50 ha of peppers were grown, while in 2005 pepper was

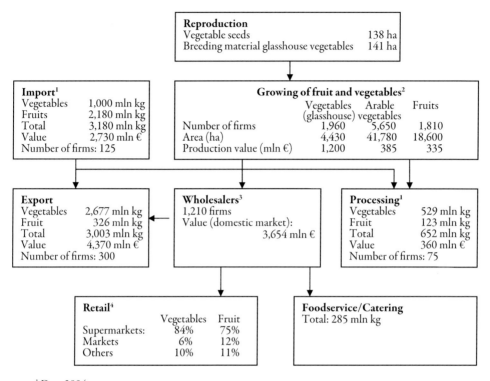

Figure 2.5. The fruit and vegetable chain in The Netherlands in 2005.

the second crop grown under glass with 1,240 ha. On a global scale The Netherlands is a small producer of tomatoes, cucumbers and peppers compared to countries such as Turkey, Egypt and Spain. Even so The Netherlands is the third most important exporter of these vegetables, after Spain and Mexico. Each year 70% of the total vegetable production under glass is exported, while for tomatoes, peppers and cucumber this percentage is even higher. Figure 2.6 shows the top three fruits and vegetables grown in The Netherlands based on their shares in the export. The total area of vegetables under glass has decreased from 5,275 ha in 1971 to 4,430 ha in 2005, while the number of firms which grow vegetables under glass has decreased from approximately 5,500 in 1971 to 1,960. Many growers of fruit and vegetables decided to grow flowers or potted plants.

The most important fruits grown in The Netherlands are apples and pears with a strong shift from growing apples to pears. The main reason is the strong international competition and the decreasing domestic consumption of apples. The planted area for apples has decreased by 49%

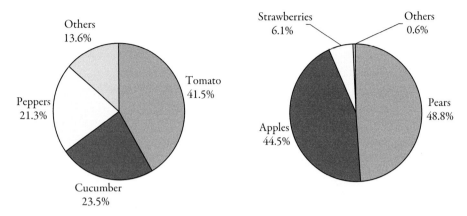

Figure 2.6. The top three Dutch fruit and vegetables as % of their respective export values in 2005 (Frugi Venta, 2006).

to just below 10,000 ha in 2005. The area of pears has increased by 30% to 6,640 ha in 2005. Furthermore, the trend for buying more processed and/or convenience food has remained. For example, the turnover of the 75 processors of fruit and vegetables was 360 million Euros in 2005, an increase of 25% compared to 2002. Moreover, re-exporting of fruits and vegetables is increasingly important for The Netherlands. In 2005 the export value of fruit and vegetables was 4.4 billion Euros, compared to 2.5 billion in 1990, while since 1990 the domestic production value of fruit and vegetables has hardly changed (2.2 billion Euros). Big wholesalers, such as the Greenery (www.thegreenery.com), Fruitmasters (www.fruitmasters.nl), ZON (www.zon-business.nl), Bakker Barendrecht (www.bakkerbarendrecht.nl) and Haluco (www.haluco.nl) play important roles in the import and export of fruit and vegetables. Although the Greenery is the most important trader on the Dutch market with a turnover of 1.9 billion Euros in 2005, it has only a market share of 5% on the European market. The activities of importers, exporters and wholesalers are highly concentrated, see Table 2.5. In 2005, the largest 6% of exporters realise 44% of the total turnover while, the largest 10% of the importers realise 55% and the largest 7% of the domestic wholesalers realise 58%, and the largest 12% of the processors realise 79% of the total turnover. Due to the large domestic production, the import of vegetables is much smaller than the import of fruits, see Figure 2.5.

The most important market for Dutch fruits and vegetables is Germany covering almost 41% of the market (see Figure 2.7). Despite the vulnerable image of Dutch fruits and vegetables, especially tomatoes, the export to Germany has increased (Van den Oever, 2005). The second important market is the United Kingdom, making up 17% of the export, followed by Belgium with almost 9% (Frugi Venta, 2006).

In Western European countries fruits and vegetables are mainly sold by supermarkets. Supermarkets sell 84% of the vegetables and 75% of the fruits on the Dutch market (see

Table 2.5. The number of firms, turnover and market share of different fruit and vegetable traders in 2004 (Frugi Venta, 2006).

Kind of firm	Number of firms	% of the total number of firms	Total turnover	Market share of largest firms
Exporters				
Turnover > 0.5 mln Euros	300	94%	4,019	56%
Turnover > 50 mln Euros	17	6%	1,752	44%
Importers				
Turnover > 0.5 mln Euros	125	90%	1,332	45%
Turnover > 25 mln Euros	13	10%	730	55%
Domestic wholesalers				
Turnover > 0.5 mln Euros	432	93%	3,654	42%
Turnover > 25 mln Euros	32	7%	2,133	58%
Processors				
Turnover > 0.5 mln Euros	75	88%	284	21%
Turnover > 7.5 mln Euros	9	12%	76	79%

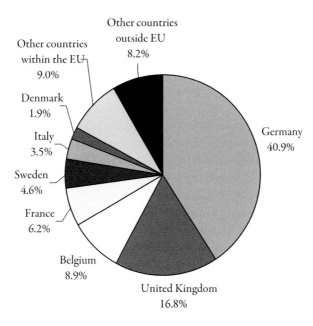

Figure 2.7. Destinations of Dutch export of fruit and vegetables in 2005 in percentages of the total export (Frugi Venta, 2006).

Figure 2.5). For supermarkets the presentation of products, assortment and reliability of delivery are important factors. Hence, adequate logistics in transporting the fruits and vegetables from the distribution centres to the retailers is crucial for success. It is interesting that on the main export market, Germany, discounters have a market share of 52% for fruits and vegetables. These German discounters buy a lot of Dutch fruit and vegetables (Berkhout and Van Bruchem, 2006).

2.4 The flower and potted plant chain

In Figure 2.8 the flower and potted plant chain is depicted with the firms and product streams, including data about the number of firms, values of the imports and exports, etc. The activities in the first step in the flower and potted plant chain, reproduction, are comparable with the fruit and vegetable chain. The Netherlands is market leader for the improvement, selection and reproduction of breeding materials. The current trend is that the management of these activities is still in the Netherlands, but a large part of the labour intensive activities are carried out in low wages countries (Wijnands and Silvis, 2000). The total production value of flower and potted plant increased to 3.7 billion Euro in 2005, an increase of 65% compared to 1990. Cut flowers make up the most important part, both in terms of number of firms and planted areas, although the share of potted plants in trading rapidly increases at the expense of flowers because (Berkhout and Van Bruchem, 2006):
1. Consumers prefer potted plants above flowers in times of economical malaise, because potted plants have a longer plant life.
2. The added value of potted plants in export is higher than for flowers, especially because of the broad assortment
3. The export of flowers suffers from strong competition from African countries.

Table 2.6 shows the share of a number of flowers and plants in the turnover of the Dutch flower auctions in 2005. Rose is still the most important flower grown in The Netherlands, with 780 ha, a share of 25% in the total area of cut flowers in 2005, although this share was 65% in 1971. For Chrysanthemum a comparable development was visible in time. The Tulip was the third flower being grown in The Netherlands. Phalaenopsis, Ficus, Dracaena, Kalanchoë and Anthurium are the most important potted plants in The Netherlands.

The Netherlands plays a very important role in the world wide trade of these products. The total export value of flowers and plants (including breeding material) was 5.1 billion Euros in 2005, the highest value ever. The export value of potted plants grew by 6.5% in 2005, twice the increase of the growth of flowers. Approximately 88% of Dutch flowers and 80% of Dutch plants are traded on the flower auctions. The main buyers on the auction are wholesalers who buy 90% to 95% of the products. They export 90% of the products and sell 10% to domestic retailers. Like most markets in agriculture, the export market for flowers and potted plants shows a very skewed distribution with regard to the number of firms and their export shares in 2005, see Table 2.7 (HBAG, 2006). In 2005 5% of the total number of exporters realised

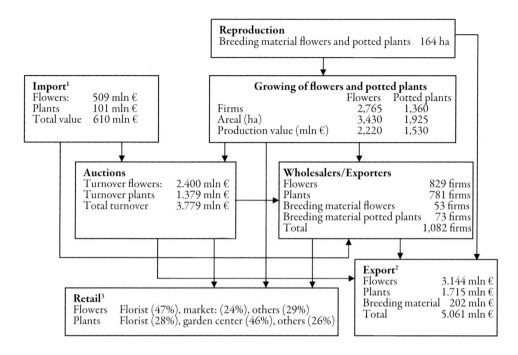

¹ Value of the import according to turnover of flower auctions
² Value according to turnover of exporters
³ Marketshare according to data of 2004
Source: BBH, CBS, HBAG, Product Board Horticulture and VBN

Figure 2.8. The flower and potted plant chain in The Netherlands.

Table 2.6. The percentages in the total turnover at the Dutch auction of the top-ten flowers and potted plants in 2005 (www.flowercouncil.org).

Plants

Kind of flowers	2005	Kind of plants	2005
Rosa	30.4%	Phalaenopsis	13.0%
Chrysanthemum (cluster)	12.2%	Dracaema	3.8%
Tulipa	8.0%	Kalanchoë	3.7%
Lilium	6.8%	Anthurium	3.6%
Gerbera	5.1%	Ficus	2.9%
Cymbidium	2.9%	Chrysanthemum	2.5%
Freesia	2.4%	Rosa	2.5%
Anthurium	1.9%	Hydrangea	2.4%
Chrysanthemum	1.7%	Spathiphyllum	2.1%
Alstroemeria	1.6%	Hedera	1.9%

half of the total turnover, see Table 2.7. Important traders of flowers and of potted plants are for example, Royal Lemkes (www.lemkes.nl), Blumex (www.blumex.nl), Intergreen (www. intergreen.nl), Kurt Schrama (www.schrama.nl) and OZ Planten (www.ozplanten.nl).

In Figure 2.9 the destinations of the Dutch flowers and potted plant export are depicted. Like in other agri-food supply chains, the most important export market is Germany with a share of almost one third. The United Kingdom comes second and is followed by France and Italy.

Dutch exporters of flowers deliver 30% to 35% of the products to wholesalers in the most important export countries. In Germany and the United Kingdom, cash and carry firms are also an important marketing channel (HBAG, 2006). As opposed to flowers, potted plants

Table 2.7. Number of exporters in 2005 according to export turnover in mln Euros (HBAG, 2006).

Turnover in 2005 (in mln Euros)	Number of firms	% of total number of firms	Turnover share
> 40	21	2%	32%
20-40	36	3%	20%
10-20	58	5%	16%
5-10	94	9%	13%
2-5	154	14%	10%
< 2	719	67%	9%
Total	1,082	100%	100%

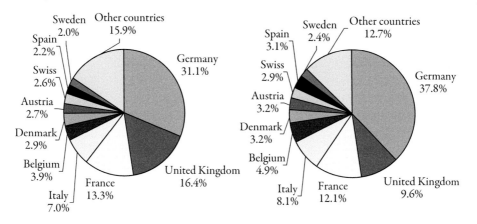

Figure 2.9. Export share (in %) of countries for Dutch flowers and potted plants in 2005 (HBAG, 2006).

are mainly sold by garden centres and building material chains with a market share of 65% to 70% in the most important export counties. In The Netherlands the florist is the main marketing channel for flowers and potted plants, but his market share is declining in favour of the supermarkets.

Interestingly, The Netherlands is also the market leader for the import of flowers and potted plants with shares of respectively 65.8% and 70.4%. Most imported flowers (65%) come from African countries, such as Kenya, Tanzania, Uganda, Ethiopia, Zambia and Zimbabwe followed by the Middle East and Latin America. Costa Rica is the most important supplier of potted plants.

In 2005 the total turnover of the flower auctions in The Netherlands was 3.8 billion Euros. The Dutch auctions are co-operatives that are meant to assist their suppliers in their commercial activities. The traditional way of working is by means of the auction clock, resembling spot-market transactions. The Dutch auction system works according to the price reduction principle in which the price is adjusted downwards until one of the traders responds. Since about ten years, the auctions have also allowed suppliers to directly sell to buyers. In these relationships the auctions act as brokers to arbitrate the sales. The latest development is that traders of flowers and potted plants increasingly develop long-term relationships with growers (Alleblas and De Groot, 2000). In these channels, often called fixed lines, the traders and growers make their own decisions about the delivery times, quantities and prices. The mediation department of the auction only handles the financial aspects of the transactions (Claro, 2003). At the moment, the majority of flowers is still sold by the auction clock, while the majority of plants is sold by mediators and fixed channels. In 2006 the two main flower auctions in The Netherlands Flora Holland and Bloemenveiling Aalsmeer intended to merge. The reason for this merger is the increased international competition. Due to the merger a network of auctions is created, providing suppliers with a more effective and efficient way of working (Flora Holland, 2007). At the moment, The Netherlands Competition Authority has approved this merger.

2.5 European food safety and quality regulations

In the Dutch agro-complex and also in the three individual chains, quality and safety issues have been extremely important, especially due to the crises that have occurred in the past. The remaining part of this chapter will especially focus on the quality and safety measures (including phytosanitary measures) that have been implemented in agri-food supply chains. This section discusses the European food safety and quality legislation and is based on Van der Meulen and Van der Velde (2004).

From the beginning of the EU in 1957 until the eruption of the BSE crisis in the mid-1990s European food law was principally directed towards the creation of an internal market for food products in the EU. The BSE -crises and other food crises (see Box 2.1.) that followed

---

Box 2.1. Overview of the most important agriculture crises.

1996: Boviene spongiforme encefalopathie (BSE) or 'Mad cow disease' turned out to be dangerous for humans. The consumption of organs of cows can result in the human variant of BSE, Creutzfeld-Jakob Disease. BSE was widely disseminated among British cows, because carcasses of sick cows have been processed in animal feed. Approximately 2.75 million cows were killed.

1997: In February 1997, **Classical Swine Fever** broke out in the South of The Netherlands and 429 firms were contaminated. In order to stop the virus the animals of 1286 firms were prevently killed, resulting in 1.8 million killed pigs.

1999: The Belgian firm Verkest mixed motor oil in fats intended for animal feed. As a result Belgian chickens got too high levels of **dioxin** in their meat and eggs. However, many of these products were already processed and retailers in many countries including The Netherlands had to withdraw many products.

2001: In the United Kingdom, **Foot and Mouth Disease** broke out and reached The Netherlands via France. In The Netherlands 26 firms were contaminated and 265.000 animals of 2600 firms had to be prevently killed. Export of agriculture products was stagnated for a long time.

2002: The Belgian Firm Bioland mixed Irish pharmaceutical waste with animal feed. As a results pigs were contaminated with **MPA hormone** and became temporal infertile. Although MPA is not hazardous for the Public Health 20.000 contaminated pigs were prevently killed. Costs in The Netherlands: 100 million Euros.

2003: Dioxin in German animal feed. The firms had also delivered feed to Dutch farmers. As a result 243 firms in The Netherlands with cattle were not allowed to provide animals to slaughterhouses

2003: Aviaire Influenza in The Netherlands. Animals from poultry farmers in the neighbourhood of contaminated firms had to be killed. The number of chickens in The Netherlands dropped from 90 million to 40 million. The sector had a 2.5 billion Euros loss of turnovers and 0.5 billion Euros loss of revenues.

2003: Aviaire Influenza ('Bird Flu') in South East Asia. The virus of Aviaire Influenza in Asia can transform into a for human hazardous variant H5N1. Later on, contaminated birds were found in many other countries over the world, including many European Countries such as Germany, France and Sweden.

2006: Bluetongue a disease for ruminants, mainly for sheep, was found in August in The Netherlands. Some days later the disease was also found in Belgium and Germany. Infected animals were vaccinated. The number of infected firms mounted 317 in October in The Netherlands.

---

soon after brought to light many serious shortcomings in the existing body of European Food Law. Public awareness of the epidemic, and the time it had taken the British and European authorities to address it, presented a major challenge to European co-operation in the area of food safety and quality. When the extent of the crises became public, the EU issued a ban on British beef exports. In response, Britain adopted a policy of non-co-operation with the European institutions and sought to deny the extent and seriousness of the BSE problem. Following these developments, the EU instituted a temporary inquiry committee to investigate the actions of the national and the European agencies involved in the crises. One of the recommendations of the inquiry committee was to improve the structure of European food law. As early as May 1997, the Commission published a Green Paper on the general principle of food law in the EU. Consumer protection was made the first and foremost priority. At the European top in Luxemburg at the end of 1997 the European Council adopted a statement on food safety. The Commission kept the pressure on beyond 1997, eventually gaining the support of the European Court for the measures taken against Britain at the height of the crisis. Meanwhile public attention had turned to a new food safety scare; the Belgian dioxin crisis. The Commission proved it had learned its lesson from its experience with BSE, and moved quickly and efficiently to protect consumers. Nonetheless, the dioxin crisis brought to light further shortcomings in European Food Law. Therefore, food safety remained a priority issue. On the 12th of January 2000 the Commission published its White Paper on food safety[3].

The White Paper aimed to restore and maintain consumer confidence. To achieve this, it proposed an ambitious legislative program. Eighty-four laws and policy initiatives were scheduled for the near future. The White Paper focused on a review of food legislation in order to make it more coherent, comprehensive and up-to-date and strengthen enforcement. Furthermore, the Commission backed the establishment of a new European Food Safety Authority (EFSA), to serve as the scientific reference point for the whole EU, and thereby contribute to a high level of consumer health protection[4]. Besides setting up the Independent Authority the White Paper called for a wide range of other measures to improve and bring coherence to the body of legislation covering all aspects of food products from 'farm to fork'. A new legal framework was proposed to cover the whole of the food chain, including animal feed production and clearly attributed primary responsibility for safe food to industry, including appropriate official controls at both national and European level. The ability to trace through the whole food supply chain was considered a key issue. Only two years after the White Paper was published, the corner stone of new European food law was laid: Regulation 178/2002. This regulation is often referred to in English as the 'General Food Law', (GFL). The main objective of the GFL is to secure a high level of protection of public health and consumer interests with regard to food products. The GFL seeks to accomplish this by three sub-objectives[5]:
1. To lay down the principle on which modern food legislation should be based in the EU as well as in the member states.

---

[3] Unlike a Green Paper that is intented mostly as a basis for public discussion, a White Paper contains concrete policy intentions.

[4] The operation of the EFSA as an independent entitiy is intended to ensure that there is a functional separation of the scientific assessment of risk from risk management decisions. The reason is that scientific risk assessment should not be swayed away by policy or other external considerations to garantuee impartiality and objectivity.

[5] These are the first three of the the the eighty-four legislative initiatives mentioned in the White Paper on food safety.

2. To establish the EFSA.
3. To establish procedures for food safety crises including the so-called Rapid Alert Systems.

It should be noted that the GFL is not a code encompassing all food legislation. It is the foundation of a general part of food law. Next to it many other European and national rules and regulations continue to play their role. According to the GFL, food law shall pursue one or more of the general objectives of a high level of protection of human life and health and the protection of consumers' interests, including fair practices in food trade, taking account of, where appropriate, the protection of animal health and welfare, plant health and the environment.

The core of the new European approach to food law is that it is based on risk analysis as a process consisting of three interconnected components, *risk assessment, risk management and risk communication*. *Risk assessment* is seen as a scientifically based process consisting of four steps: hazard identification, hazard characterisation, exposure assessment and risk characterisation. *Risk management* is a broad concept. It means the process of weighting policy alternatives in consultation with interested parties, considering risk assessment and other legitimate factors, and, if needed, selecting appropriate prevention and control options. *Risk communication* means the interactive exchange of information and opinions throughout the risk analysis process. This includes hazards and risks in themselves, as well as risk related factors and perceptions of risk among assessors, managers, consumers, feed and food businesses, the academic community and other interested parties. To a large extent risk management and - communication is the domain of the European Commission and the national authorities.

European food safety requirements apply to imported food as well. These requirements on food safety not only address the state of the product when it arrives at the European border, but also the way it has been handled in processing and trade. This principle therefore implies considerable extra-territorial ambition of EU food law. Exported food must also comply with the European food safety standards. This principle on the one hand makes the European origin of a food product a quality guarantee. On the other hand it facilitates controls and enforcement as all production in the EU in principle has to meet the same safety standards, regardless of the market for which they are producing.

The EU legislation of food also takes into account legislation from other international official regulatory bodies. On the global level the World Trade Organisation (WTO) tries to remove barriers to trade. To achieve this, several measures have been taken. First tariff barriers were reduced and to the extent that this was successful non-tariff barriers became more a concern. In food trade differences in technical standards like packaging requirements may cause many problems, but mostly concerns about food safety, human health, animal and plant health induces national authorities to take measures which hampers the free flow of trade. To address these concerns two WTO treaties were made: the *Agreement on the Application of Sanitary and Phytosanitary Measures (the SPS Agreement)* and the *Agreement on Technical Barriers to Trade (the TBT Agreement).*

The *SPS Agreement* recognises protection of human, animal and plant health and life by means of sanitary and phytosanitary measures. The basic aim of the SPS agreement is to maintain the sovereign right of any government to provide the level of health protection it deems appropriate and to ensure that these sovereign rights are not misused for protectionist purposes and do not result in unnecessary barriers to international trade. Therefore, the measures must be scientifically justified and they may not be discriminating, nor constitute disguised barriers to international trade. The most important standards on food safety are to be found in the *Codex Alimentarius* (described in Box 2.2). However, if the measures are in conformity with international standards, no scientific proof of their necessity is required.

Box 2.2. The Codex Alimentarius.

The *Codex Alimentarius Commission, CAC* (www.codexaliamentarius.net), which was created in 1961 by the Food and Agriculture Organisation (FAO) and the World Health Organisation (WHO) develops food standards, guidelines and related codes of practice under the Joint FAO/WHO Food Standards Programme. About 165 countries, representing 98% of the world's population participate in the CAC. The Codex Alimentarius can be seen as a book filled with food standards. Besides the food standards, Codex Alimentarius includes advisory provisions called codes of practices or guidelines. At present the Codex comprises more than 240 commodity standards, over 40 food hygiene and technological codes of practice; over 1,000 food additives and contaminants evaluations and over 3,200 maximum residue limits for pesticides and veterinary drugs. Finally, the Codex Alimentarius includes requirements on labelling and presentation and on methods of analysis and sampling. The inclusion of the Codex Alimentarius in the SPS Agreement, greatly enhances its significance.

The importance of the Codex Alimentarius is also recognised in the GFL, because it introduces an obligation to take international standards like the CAC into account. In addition, the Codex Standards help to define the limits that European law sets to national legislators. If requirements of European food safety law are not in conformity with the Codex Alimentarius, sooner or later, they will be contested under the SPS Agreement as barriers to international trade. For the food sector the practical result is that they have access to the majority of the world's markets if their products are up to Codex standard.

The *TBT Agreement* covers all technical requirements and standards, such as labelling, which are not covered by the SPS Agreement. Therefore, the SPS and the TBT Agreement can be seen as complementing each other.

As a result of the crises, the EU has clearly recognised, that the safety of food depends largely on the way is has been produced. For this reason regulations have been made to ensure that safe methods of production are used. On the 1st of January 2006 Regulation 852/2004 came into force. The regulation imposes a general obligation on food business operators to ensure that in all stages of production, processing and distribution food under their control satisfies the relevant hygiene regulations. At the heart of these requirements is the Hazard Analysis of Critical Control Points (HACCP) system (see Box 2.3 for more information about HACCP).

Box 2.3. The HACCP approach.

HACCP

In many countries world-wide, legislation and private quality management systems on safety requires HACCP. It is a system to assess hazards and establish control systems that focuses on prevention rather than relying on end-product testing. It is a system based on seven principles. Each firm has to apply these principles to its own specific situation, because it is not a tangible manual with prescriptive actions that can be directly applied to firms. In the table below, the 7 principles of HACCP are shown.

Guidelines for implementing (FAO/WHO,1997).
7 HACCP principles:
1. Conduct a hazard analysis
2. Determine the Critical Control Points
3. Establish critical limits
4. Establish a system to monitor
5. Establish the corrective action to be taken when monitoring indicates that a particular CCP is not under control
6. Establish procedures for verification to confirm that the HACCP system is working effectively
7. Establish documentation concerning all procedures and records appropriate to these principles and their application.

The Codex Alimentarius is a strong supporter of HACCP and provides codex on how HACCP can be applied in different sectors.

The formulation of the specific standards that must be adhered to is left to industry. Moreover, the GFL and the new hygiene regulation require that food and food ingredients are traceable. Food firms must keep comprehensive records of exactly where their food materials originated from and where it went. However, this regulation does not require an intact paper trail to accompany each individual food ingredient from farm-to-fork. Traceability requirements go only one step up and one step down the food chain. In addition, Regulation EG 853/2004[6]

---

[6] http://eur-lex.europa.eu/LexUriServ/site/nl/consleg/2004/R/02004R0853-20060101-nl.pdf

which includes a number of specific hygiene requirements for food of animal origin and Regulation EG 854/2004[7] that prescribes the official way of inspecting products of animal origin came into force at the same date.

2.6 Private European quality management systems

Besides the public regulations for assuring quality and safety in agri-food supply chains, many private quality management systems exist on which this section will focus. Recently, particularly big European retailers have developed initiatives to commit their suppliers to strict food quality management systems. Only a few years ago, food safety and quality were not important issues for retailers. At that time, retailers did not have food safety and quality programs or a quality department or adviser. This has changed dramatically and many retailers have developed quality management systems for their suppliers. These standards contain comprehensive norms with regard to food safety, product and process management and hygiene of the personnel. Retailers expect legal, technical and financial advantages from these systems. The standards were developed to assist retailers in their fulfilment of legal obligations and protection of the consumers, but include nowadays even more stringent requirements to food safety and quality than the legislative demands (Havinga, 2006). In Table 2.8 a number of retail initiated quality management systems are described. Since many of the discussed systems are based on HACCP and ISO, with respect to hygiene and food safety concerns, the various quality systems have similar features to a large extent. Although there are acknowledgements of some systems, it seems that there is still a long way to go to harmonise the different systems.

With regard to the standard aimed at the direct suppliers of retailers, such as food manufacturers, processors and traders, the Global Food Safety Initiative, GFSI, (www.ciesnet.com) has been established to harmonise the various standards on a global scale. As a result suppliers will be treated more equally throughout the world. At the moment, the supermarket chains work with different standards. Some use firm-specific standards, whereas others support a retail standard (for example BRC, IFS, Dutch HACCP, SQF2000). There are also retailers who still have their own auditing standard in addition to a GFSI recognised standard. For example, the British retailer Tesco accepts BRC and would also accept IFS, but still will undertake its own audits with their Tesco checklist. A special note is reserved for the French and Belgian retailers: most of the French retail firms who are members of the FCD (Federation du Commerce et Distribution, www.fcd.asso.fr), have participated in the IFS working group. However, each individual French retailer will make its own policy towards accepting IFS audit reports. The Belgium retail federation FEDIS has decided that their members will accept all GFSI recognised standards. FEDIS has developed a standard for small and medium sized firms, which can be downloaded from their site (www.fedis.be). However, this standard is not benchmarked under GFSI (www.evmi.nl).

---

[7] http://eur-lex.europa.eu/LexUriServ/site/nl/oj/2004/l_226/l_22620040625nl00830127.pdf

Table 2.8. International quality management systems in agri-food supply chains.

System	Aimed at	Aim/set up of the system
British Retail Consortium Standard (BRC)	Firms supplying retail branded food products. Certification related to the product and process conditions	BRC requires having a quality system operational, and a HACCP plan including environmental issues, product process and staff. Consist of 10 fundamental criteria, 42 statements of intent criteria and 227 requirements (no different levels anymore).
International Food Standard (IFS)	Audit standard for retailer/and wholesale branded food products, including tobacco producers and distribution. Certification related to the product and process conditions	To ensure food safety and the quality level of producers of retailer branded food products. Two levels: basic (minimum requirements) and advanced level (higher standards). Recommendations are included for the requirements to obtain an outstanding position in the advanced level (best practice in industry). 4 critical knock-out criteria, 225 foundation level criteria, 60 higher level criteria and 46 recommendations
Safe Quality Food (SQF)	Standard for suppliers of raw materials, ingredients, food products, beverages and services	SQF system addresses food safety and quality, but also other issues such as animal welfare, environmental impact, ethical production, organic production and religious preparation requirements can be included. Three levels, increasing in their strictness: Level 1: Food safety fundamentals; Level 2: Accredited HACCP Food Safety Plans; Level 3: Comprehensive Food Safety and Quality Management System Development
Dutch HACCP	Requirements for a *certified* HACCP based Food Safety System. Management standard for the primary sector, processing industry, distribution and logistics	One level: all requirements of the standard and of the prerequisite program have to be met. Total amount of criteria: 138.
International Standard Organisation (ISO 22000)	Management standard for any organisation in the food chain, including feed producers and service providers	One level: all requirements of the standard, including a prerequisite program have to be met. Amount of criteria: 200
Euro-Retailer Produce Good Agricultural Practices (Eurep-GAP)	Primary producers in agri-food supply chains	Eurep-GAP supports the use of HACCP and members are obligated to comply with inter (national) legislation. They have to show commitment to issues such as reduction of environmental damage, pesticide use, efficient use of natural resources, health for safety for employers and traceability efforts.
Qualität und Sicherheit (QS)	All stages in multiple food supply chains	Makes sure that firms fulfil the legal requirements and food safety criteria that go beyond legal regulations.

Sanction possibilities	Frequency of auditing	Based on	Origin
Three categories of sanctions in case shortcomings: critical (blockade), major (improvement report), minor (recommendations).	Every 6 or 12 months	HACCP, ISO	British retailers
The auditor works through a detailed checklist and gives a rating of the various elements. If the scores are too low for a certain level corrective actions have to be performed and a new audit is necessary. Otherwise the certificate will be suspended.	Once a year for the basic level and once per 18 months for the higher level. Seasonal products: once a year	HACCP, ISO, BRC	German, French and Swiss retailers (German/French BRC equivalent)
Four sanctions: observation, simply suggestions to improve a firm's processes. Minor non-conformance, has to be improved before next audit or earlier. Major non-conformance, a time limit will be set to the firm to correct the problem. Critical non-conformance, immediately a corrective action is needed.	Twice a year. If in 3 years no major or critical incident arises, the audit frequency becomes once a year	HACCP and ISO	Australian retailers, but has been applied widely in other regions
Sanctions varying from temporal blockage and recall to recall and destruction of the product. Firms have to adjust processes in case of non-confirmation.	Twice a year	HACCP	Dutch retailers (Foundation for Certification Food Safety)
Corrective actions and in case of repetitive reminders in case of non-confirmation, withdrawal of the certificate	Usually twice a year; sometimes once year	HACCP	International Standard Organisation (ISO)
Three sanctions, recommendations, minor musts and major musts. When a less than 95% of the minors must is fulfilled or one major must, the certificate will be temporarily suspended, in case of non-improvement permanently.	At least one announced audit per year in addition with unannounced audits	HACCP	Many European retailers
Follow-up audit, higher audit frequency, warning, fine or exclusion from the system	The better the results the longer the audit interval	Eurep-GAP, IKB	German retailers

The GFSI has benchmarked four international food safety standards for food manufacturers, processors and traders: BRC, the Dutch HACCP Code, IFS and the SQF 2000 code. For this, the GFSI has developed a Guidance Document which contains three key elements: Food Safety Management Systems, Good Practices for Agriculture, Manufacturing or Distribution, and HACCP. Once a food standard has been benchmarked successfully, the standard is 'acknowledged'. The benchmarked food safety standards can be applied by food suppliers throughout the supply chain, upon agreement with retailers, when defining contracts for sourcing of products. The application of the benchmarked standards to particular products will be at the discretion of retailers and suppliers. This process will vary in different parts of the world, depending on firm policies, general regulatory requirements, product liability and due diligence regulations. Figure 2.10 illustrates the acceptances of standards by a number of important retailers in a number of European countries (not exhaustive). Retailers in the grey coloured countries do not ask for any of the standards within GSFI. Instead they use firm specific quality management standards (www.evmi.nl).

ISO 22000 (www.iso.org) has not been benchmarked by GFSI, simply because ISO has not asked for a benchmark operation. Even if ISO would, ISO 22000 would not comply with GFSI's demands, because the standard does not require on-site inspections. Most importantly, the ISO norm requires the implementation of Good Manufacturing Practices (GMP) but does not contain specific demands in terms of GMP.

Furthermore, the system Eurep-GAP (www.eurepgap.org) is especially aimed at the primary producers and not to the direct suppliers of retailers. Eurep-GAP stands for Euro-Retailer Produce Good Agricultural Practices working group and is a platform in which the major European food retailers are grouped. The system Eurep-GAP was firstly introduced and fully developed in the fruit and vegetable chain, but was later on expanded to other sectors, like flowers and ornamentals, meat and fish (Van Plaggenhoef *et al.*, 2003; Bondt *et al.*, 2005).

Eurep-GAP supports the principles of and encourages the use of HACCP, but takes the reduction of environmental damage into account such as the pesticide reduction, efficient use of natural resources and health and safety for employers (Van Plaggenhoef *et al.*, 2003; Havinga, 2006). In the poultry meat chain and the flower and potted plant chain, systems such as respectively IKB and MPS (see Section 2.8.4. and 2.10) are in a process of acknowledgement as equivalents of the Eurep-GAP systems. Since March 2007, firms that participate in Eurep-GAP will be certified according to Eurep-GAP Integrated Farm Assurance, which means that the separated quality modules (for meat or fruit) have been integrated in one system, resulting in more transparency and lower costs (www.eurep.nl).

Qualität und Sicherheit' GmbH, QS, (www.q-s.info) is a merger of six shareholders in the German agriculture, including major (especially German) retailers, such as Metro, Edeka, Rewe, Kaiser's Tengelmann, Aldi, Coop, Globus, Kaufland, Marktkauf and Wal-Mart. QS is internationally active and already works with different standards used in neighbouring

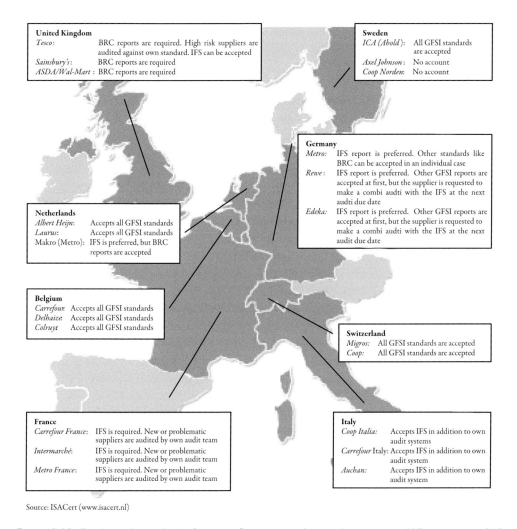

United Kingdom
Tesco: BRC reports are required. High risk suppliers are
                  audited against own standard. IFS can be accepted
Sainsbury's: BRC reports are required
ASDA/Wal-Mart : BRC reports are required

Sweden
ICA (Ahold):   All GFSI standards
                 are accepted
Axel Johnson : No account
Coop Norden: No account

Germany
Metro: IFS report is preferred. Other standards like
           BRC can be accepted in an individual case
Rewe :   IFS report is preferred. Other GFSI reports are
           accepted at first, but the supplier is requested to
           make a combi audti with the IFS at the next
           audti due date
Edeka: IFS report is preferred. Other GFSI reports are
           accepted at first, but the supplier is requested to
           make a combi audti with the IFS at the next
           audit due date

Netherlands
Albert Heijn: Accepts all GFSI standards
Laurus: Accepts all GFSI standards
Makro (Metro): IFS is preferred, but BRC
                   reports are accepted

Belgium
Carrefour. Accepts all GFSI standards
Delhaize: Accepts all GFSI standards
Colruyt   Accepts all GFSI standards

Switzerland
Migros: All GFSI standards are accepted
Coop: All GFSI standards are accepted

France
Carrefour France: IFS is required. New or problematic
                     suppliers are audited by own audit team
Intermarché: IFS is required. New or problematic
                     suppliers are audited by own audit team
Metro France: IFS is required. New or problematic
                     suppliers are audited by own audit team

Italy
Coop Italia: Accepts IFS in addition to own
                    audit systems
Carrefour Italy: Accepts IFS in addition to own
                    audit system
Auchan: Accepts IFS in addition to own
                    audit system

Source: ISACert (www.isacert.nl)

Figure 2.10. Food retail standards for manufactures, traders and processors: Who accepts which standard? (Source: ISACert, www.isacert.nl)

countries. For example, firms complying with the IKB system in the poultry meat chain can participate in QS. The difference with the other quality management systems summarised in Table 2.8 is that QS is oriented at the total supply chain, whereas the other systems, except Eurep-GAP are mainly limited to the direct suppliers of retailers or to primary producers (like Eurep-GAP).

2.7 National food safety and quality regulations

The investigated chains in the present study have also to comply with Dutch quality and safety legislation. Most Dutch laws and regulations are derived from the regulations of the EU or are implementations of these regulations. The most important Dutch laws regarding quality and safety of food and feed is the *Dutch Food and Non-Food Law* and the *Agricultural Quality Law* and are discussed below. Other important laws are for example, the *Meat Inspection Law*, the *Animal Welfare and Health Law* and the *Pesticide Law*. These laws are discussed in the sections describing the public and private inspections strategies in each chain.

The main principle of the Dutch *Food and Non-Food Law*[8] (in Dutch: Warenwet) is simply that food has to be safe. In addition to food products themselves, the Dutch Food and Non-Food Law concentrates on the raw materials, production processes and selling points (Van Plaggenhoef and Batterink, 2003). For example, important issues such as the number of pathogenetic micro-organisms and the use of decontamination stuffs such as processing materials are incorporated in this regulation. This law also has regulations on the labelling of food products. It states that the producers and or traders of food products have the primary responsibility for the safety of food. Since 1996, this law further has obliged food-processing firms to use a HACCP-based system for the control of product (food) safety. Therefore, the new regulation of the EU 852/2004 that came into force in January 2006 (see also Section 2.5.) which requires that all firms involved in food production have to work according to the HACCP did not change much for the Dutch agri-food supply chains, because this requirement was already included in the Food and Non-Food Law. The Food and Consumer Product Authority (VWA, see Box 2.4 for more information) is primarily responsible for the inspection of this law. This law is a framework and consists of a general part on which regulations can be attached in order to keep the law up-to-date without many difficult legal procedures. An overview of all regulations for this law is available at http://wetten.overheid.nl. For up-dating the Food and Non-Food Law, the government consults the Regulier Overleg Warenwet (in English: Regular Consult Food and Non-Food Law) four times a year. This is a panel including representatives from food manufacturers, food traders, Product Boards and consumers.

The Ministry of Agriculture, Nature and Food Quality, the Ministry of Health, Welfare and Sport and the VWA are also involved. When the government decides to implement regulations that are not confirmed by this panel, they have to motivate this in the instructions of the legal text. More information can be found at (http://www.row.minvws.nl/).

Another important law applicable to multiple chains in agriculture is The *Agricultural Quality Law*[9] (in Dutch: Landbouwkwaliteitswet) which contains regulations concerning the origin and the labelling of raw materials for food products that originate form agricultural products.

---

[8] http://wetten.overheid.nl/cgi-bin/deeplink/law1/title=Warenwet

[9] http://wetten.overheid.nl/cgi-bin/deeplink/law1/title=Landbouwwet

Box 2.4. The VWA, the AID and the Inspection Desk Agriculture.

VWA (www.vwa.nl)

The Food and Consumer Product Safety Authority (in Dutch: Voedsel en Waren Autoriteit, VWA) was set up on the 10th of July 2002. Because of a number of food safety crises the Dutch Parliament felt the need for a strong organisation to protect safety of food. In addition, developments in the international sphere (such as the GFL) demanded for one national authority, responsible for inspection, risk assessment and risk communication. Two organisations, the Inspectorate for Health Protection and Veterinary Public Health (in Dutch: Keuringsdienst van Waren, KvW) and the National Inspection Service for Livestock and Meat (in Dutch: Rijksdienst voor keuring van vee en vlees, RVV), were brought under the umbrella of the VWA. The task of the VWA is to protect public health and animal health and welfare (www.vwa.nl). The VWA controls the whole production chain, from raw materials and processing aids to end products and consumption by means of a risk-based system, incorporating the appropriate supervisory arrangements. With regard to the new policy of 'Control-on-control' the VWA has started a number of pilot projects in which the own inspection tasks are aligned with the activities of private organisations(AID, 2006).

AID (www.aid.nl)

General Inspection Service (in Dutch Algemene Inspectie Dienst, AID) of the Ministry of Agriculture, Nature and Food Quality deals with issues such as the use of animal medicine, identification and registration of animals, use of animal feed, correct use of pesticides. The AID collaborates with other inspections agencies such as the VWA, the police, the labour inspection and the water management agencies. Besides inspection, the AID has tasks with regard to the detection of fraud, verification of data supplied by firms for example, to participate in a certain subsidy regulation, enforcement and communication of regulations and has therefore a somewhat broader scope of inspection topics than the VWA (www.aid.nl). The AID especially supervises the primary producers, whereas the VWA supervises firms further downstream in the chain, such as traders and processors and further upstream such as feed processors. Furthermore, the AID functions a juridical agency for the VWA.

Inspection Desk Agriculture (www.inspectieloket.nl)

Since January 2007 the Inspection Desk Agriculture (www.inspectieloket.nl) has been established in which multiple governmental agencies such as the VWA, the Labour Inspection agency, the Plantenziektenkundige Dienst (in English: Plant Disease Agency) and the Dienst Landelijk Gebied (in English: Agency for Rural Areas) work together. The Inspection Desk facilitates the exchange of planning data and the alignment of the inspections, and if possible, combines multiple inspections. The objective is to reduce the burdens of inspections for firms, which seems to be successful. In order to facilitate a proper enforcement of regulations, governmental agencies have several sanction possibilities such as warnings, fines, report on offence, administrative fines, and closure of the firm.

Its main aim is to improve the quality of agricultural products. The main focus of the law is on producers and traders of agricultural export products. Both the VWA and the AID are responsible for the governmental supervision of the law. However, in many cases inspections are executed by organisations in the sectors concerned (for example, the KCB in the fruit and vegetables sector, see Box 2.5). A significant number of the regulations in this law are dedicated to organic production.

Most inspections of the governmental agencies are based on risk analyses. For example, the AID and VWA identify each year a number of topics (e.g. animal welfare, medicine use, pesticide use, etc.) on which the inspections will focus. This means in practice that a planning is made of each topic including the number of inspection hours and the costs. However, in case of crises this planning is adapted. Examples of the number of hours planned and realised for certain inspections can be found in the yearly reports of the AID (AID, 2007). For choosing the firms a risk based approach is also followed. For example, the VWA classifies firms into a pyramid of three categories with regard to Public Health (VWA, 2006). Firms in the green basis work in a safe and responsible way, firms in the orange level in between work less safe and the firms in the red top are performing badly. The VWA applies this classification in all of its domains and works according to a bonus-malus principle, good performing firms get fewer inspection and bad performing firms get more inspections. The sanctions are also higher for firms that keep performing badly (VWA, 2007a).

Product Boards (which represent the common interests of a sector) and Industry Boards (which represent the common interests of a certain group with a comparable profession, for example, traders in fruit and vegetables) play an important role in the implementation of legislative requirements in agri-food sectors and are characteristic of Dutch agriculture. These organisations can be authorised by the government to apply and enforce regulations smoothly in their sectors. Product Boards and Industry Boards are financed by means of fees from the firms and financial compensations from the government for the work performed. Important Product Boards are the Product Boards for Livestock, Meat and Eggs (www.pve.nl) and the Product Board Horticulture (www.tuinbouw.nl). For their members, these organisations develop hygiene codes in order to comply with the legislative HACCP demand. Each three or five years the hygiene codes are evaluated in which the level of applicability and usefulness are evaluated.

In the sections below, the state-of-the-art of the public and private quality regulations including the inspection strategies are discussed in the three individual chains. It should be noted that only the most important regulations are discussed and that it is not the intention to give an exhaustive overview.

2.8 Regulations in the poultry meat chain

In this section the most important public regulations with regard to food quality and safety for the primary producers and the processors in the poultry meat chain are discussed. Furthermore, the inspections situations for both groups of firms are described. Finally, private initiatives for the assurance of quality and safety such as the Action Plan Salmonella and Campylobacter and IKB are described.

2.8.1 Public regulations

Firms in the poultry meat chain have to comply with a large number of regulations dealing with food quality and safety. The most important and well-known laws and regulations are briefly described below:

- The *Kaderwet Animal feed*[10] describes the ingredients that are allowed to be used in animal feed. For example, it is not allowed to process animal slaughter by-products in animal feed for cows, pigs, sheep and goats. Furthermore, this law describes the tolerances for a number of stuffs that are harmful when these tolerances are exceeded. Another important requirement is that feed processors and transporters are registered, which is useful in times of crisis. Finally, this law regulates the intervention of the government in case of crisis. The AID supervises this law in the primary sectors, whereas the VWA supervises this law at processors and transporters.
- The *Animal Medicine Law*[11] describes which animal medicines are allowed to be used and how they should be used both for curing animals and for the processing of feed in which medicines are mixed for preventive curing. Only registered medicines may be used in such a way that it is safe for the animals, the environment and the consumer. Regarding this, the use of growth stimulating medicines is not allowed. The AID is responsible for supervising the Animal Medicine Law.
- The *Animal Health and Welfare Law*[12] (which is relatively new) provides regulations concerning accommodation, transport and treatment of animals and the prevention and curing of animal diseases. Animal diseases can be infectious for humans or can influence the quality of animal products from animal origin as well. This law also contains requirements for the housing of animals in the intensive livestock. The VWA and the AID are responsible for the inspections.
- The *Meat Control Law*[13] includes quality and safety standards regarding animal diseases for the slaughtering process and the resulting products. This law requires that animals are inspected by a qualified veterinarian to identify possible animal disease symptoms before and after the slaughtering process. When animal diseases are identified that can be harmful

---

[10] http://wetten.overheid.nl/cgi-bin/deeplink/law1/title=Kaderwet%20diervoeders

[11] http://wetten.overheid.nl/cgi-bin/deeplink/law1/title=Diergeneesmiddelenwet

[12] http://wetten.overheid.nl/cgi-bin/deeplink/law1/title=Gezondheids-%20en%20welzijnswet%20voor%20dieren

[13] http://wetten.overheid.nl/cgi-bin/deeplink/law1/title=Regeling%20vleeskeuring

for consumer health, the meat has to be rejected and destroyed. The VWA is responsible for the supervision of this law.

- The *Destruction Law*[14] regulates the transport and the processing of animal materials that are not used for consumption, such as animals which have died on farms or slaughter waste. Such animal material could be harmful for the human health or could cause animal diseases. Therefore, the material has to be destroyed. The Destruction Law is crucial in the fight against BSE. The VWA is responsible for the enforcement of this law.

2.8.2 Inspection situation for poultry farmers

In Table 2.9 the inspection situation for poultry meat farmers is summarised with regard to quality and safety. From this overview, it becomes clear that poultry farmers have to comply with more legislative requirements than described above. More detailed information about the remaining regulations is found on the URLs provided. It is interesting to note from this overview that when for example, the number of firms inspected is only 5%, this means that an average firm has the chance that it will be inspected once in twenty years.

2.8.3 Inspection situation for slaughterhouses and cutters

The inspections for poultry slaughterhouses and cutters are conducted by the VWA and consist of *permanent inspections*, *system audits*, *system inspections* and *sampling* (VWA, 2007b).

Permanent inspections

In the slaughterhouses an Ante-Mortem inspection is mandatory. This means that a veterinarian inspects animal welfare aspects and the animal health status (as described in the Animal Health and Animal Welfare Law[15] and the Regulation on meat inspection[16]). When the Ante Mortem inspections do not identify problems, the slaughtering process can be started. Furthermore, there is Post Mortem inspection by the same veterinarian. In the poultry meat chain, the slaughterhouse carry out the Ante Mortem and Post Mortem inspection themselves, by means of private inspectors and not by means of the official veterinarian (which is a difference with other slaughterhouses) The VWA supervises these private inspectors of the poultry slaughterhouses.

---

[14] http://wetten.overheid.nl/cgi-bin/deeplink/law1/title=Destructiewet

[15] http://wetten.overheid.nl/cgi-bin/deeplink/law1/title=Gezondheids-%20en%20welzijnswet%20voor%20dieren

[16] http://wetten.overheid.nl/cgi-bin/deeplink/law1/title=Regeling%20vleeskeuring

Table 2.9. The inspection situation for poultry farmers (adapted from www.inspectieloket.nl).

Inspected issues	Aim of the inspections	Which inspection tasks are conducted	Duration inspection (hours)	Firms inspected firms (%)	Regulations addressed ***	Inspection agency
Requirements with regard to manure production and - use	To promote sustainable production, such as higher quality of water with regard to nitrogen and phosphate concentration	Inspection of administrative obligations with regard to manure production and the requirements for the supply and removal of manure	4	—	1	AID
Animal welfare	To ensure a sufficient level of animal welfare	Inspection of the requirements on animal welfare in the regulation on the production animals	4	5%	2, 3	AID
Time of beak cutting at hatcheries and rearing firms	To enforce compliance with the ban on beck cutting ten days after birth	Inspection on the way the becks are cut	2	1%	2, 4	AID
Killing of animals outside the slaughterhouse	To kill animals on a correct way	Inspection of the welfare requirements with regard of killing animals	1	<1%	2, 5, 6, 7	AID
Notification, covering, availability, and cooling of destruction materials	To enforce compliance with the requirement to notify carcasses and to prevent that risk materials are used for human consumption	Inspection of the requirements mentioned under 'inspected issues'	1.5	5%	8, 9	AID
Compliance with the ban on prohibited stuffs	To prevent and resist the use of prohibited stuffs for food producing animals	Inspection on the presence of prohibited stuffs, sampling of animals, feed and or products	6	<1%	10, 11, 12, 13	AID
Compliance with the right use and dosage of animal medicines and raw materials with a pharmaceutical effect	To promote the correct use of animal medicines and raw materials with a pharmaceutical effect	Inspection on animal medicines and the registration of them	6	5%	11	AID

Table 2.9. Continued.

Inspected issues	Aim of the inspections	Which inspection tasks are conducted	Duration inspection (hours)	Firms inspected firms (%)	Regulations addressed ***	Inspection agency
The use of animal feed, preparations and additional animal feed	To promote the correct compliance of the ban to make, use, store, process, transport, etc of feed that are not healthy, sound, safe or below minimum trade quality requirements	Inspection on the use of animal feed, preparations or additional animal feeds	2.5	5%	14, 15, 16, 17, 18	AID
The presence of hormones/residuals in animals that are intended for human consumption	To prevent that consumers are provided with food that contains too high levels of harmful stuffs	Taking samples of for example, animal urine and or animal hairs	1	7%	7, 13, 16, 18, 19	VWA
Presence of prohibited stuffs in feed	The use of pure raw materials by means of contamination analysis of study in which the use of by products becomes transparently	Taking samples of the feed that is present on the firms	1	12%	14, 17, 18	VWA
Health states of the animals	Prevention and fighting of animal diseases (especially for the ones with a notification obligation)	Sampling, control on identification and registration, veterinarian judgement, inspection of the registration with regard to the health status of animals	5	*	2, 20	VWA
To kill animals for slaughtering purposes on the primary firm	To kill animal according to animal welfare requirements and inspection of the quality of products for human consumption	Inspection of the right anaesthetise time and the minimal bleeding time	1%	**	2, 5, 6, 7	VWA

Table 2.9. Continued.

* Depends on the presence of animal diseases in The Netherlands.

** Is carried out during other inspections.

*** Regulations addressed:

1. Manure Law: http://wetten.overheid.nl/cgi-bin/deeplink/law1/title=Meststoffenwet

2. Animal Health and Welfare Law: http://wetten.overheid.nl/cgi-bin/deeplink/law1/title=Gezondheids-%20en%20welzijnswet%20voor%20dieren

3. Decision on animal welfare of production animals: http://wetten.overheid.nl/cgi-bin/deeplink/law1/title=Besluit%20welzijn%20productiedieren

4. Decision on surgery: http://wetten.overheid.nl/cgi-bin/deeplink/law1/title=Ingrepenbesluit

5. Regulation on meat inspection: http://wetten.overheid.nl/cgi-bin/deeplink/law1/title=Regeling%20vleeskeuring

6. Decision on the killing of animals: http://wetten.overheid.nl/cgi-bin/deeplink/law1/title=Besluit%20doden%20van%20dieren

7. Regulation on the hygiene of food of animal origin: http://eur-lex.europa.eu/LexUriServ/site/nl/consleg/2004/R/02004R0853-20060101-nl.pdf

8. Destruction Law: http://wetten.overheid.nl/cgi-bin/deeplink/law1/title=Destructiewet

9. Regulation animal by-products: http://wetten.overheid.nl/cgi-bin/deeplink/law1/title=Regeling%20dierlijke%20bijproducten

10. Agricultural Law: http://wetten.overheid.nl/cgi-bin/deeplink/law1/title=Landbouwwet

11. Animal medicine law: http://wetten.overheid.nl/cgi-bin/deeplink/law1/title=Diergeneesmiddelenwet

12. Regulation inspection measures with regard to certain stuffs and their residuals in living animals and products of them: http://eur-lex.europa.eu/LexUriServ/site/nl/consleg/1996/L/01996L0023-20030605-nl.pdf

13. Regulation ban on the use of certain stuffs with a hormonal working and certain stuffs with thyreostatic working and agonistics: http://eur-lex.europa.eu/LexUriServ/site/nl/consleg/1996/L/01996L0022-20031014-nl.pdf

14. Regulatory framework animal feed: http://wetten.overheid.nl/cgi-bin/deeplink/law1/title=Kaderwet%20diervoeders

15. Food and Non Food Law: http://wetten.overheid.nl/cgi-bin/deeplink/law1/title=Warenwet:

16. General Food Law: http://eur-lex.europa.eu/LexUriServ/site/nl/consleg/2002/R/02002R0178-20031001-nl.pdf

17. Regulation on feed hygiene: http://eur-lex.europa.eu/LexUriServ/LexUriServ.do?uri=CELEX:32005R0183:NL:HTML

18. Regulation on inspection of feed/feed: http://eur-lex.europa.eu/LexUriServ/site/nl/consleg/2004/R/02004R0882-20060525-nl.pdf

19. Regulation on the organisation of official inspections of products of animal origin intended for human consumption: http://eur-lex.europa.eu/LexUriServ/LexUriServ.do?uri=CELEX:32004R0854R%2801%29:NL:HTML

20. Regulation on veterinary and zoonose inspections in intracommunautary trading: http://eur-lex.europa.eu/LexUriServ/site/nl/consleg/1990/L/01990L0425-19901019-nl.pdf

System audits

During the system audits firms are inspected on the presence of a HACCP system. For slaughterhouses and cutters it is also permissible to work according to hygiene codes. If firms comply with all the requirements of the system, they will get 1 VWA audit each year. The issues that are controlled in the HACCP plan are described in Regulation EU 852/2004[17], EU 853/2004[18], EU 854/2004[19] and the Food and Non-Food Law[20]. The basic requirements of good hygiene practices deal at least with:
- Inspection of the information about the supply chain
- The design and the maintenance of the buildings
- The hygiene before, during and after the slaughter process
- Personal hygiene
- The trainings about the themes hygiene and working practices
- Vermin fighting
- The water quality
- The temperature control
- Inspection of foods that come in and leave the firms and the accompanying documentation

System inspections

Each year a slaughterhouse receives six system inspections which inspect how parts of systems for the assurance of the safety and quality of food are implemented in the firm. At least one of these system inspections deals with the basic hygiene requirements as described above and one inspection is aimed at animal by-products. The issues of the remaining inspections are free to choose for the inspectors. The choice for the issues is based on the firm specific circumstances. These inspections last approximately three hours.

Sampling

Once a year each slaughterhouse or cutter is inspected by the VWA for the regulation of microbiological criteria[21]. During this inspection many samples are taken at the slaughterhouse in order to judge compliance with microbiological criteria, this lasts three hours. Besides the burdens caused by the legislative demands, there are many other indirect burdens related to quality management, such as developing and maintaining the HACCP systems, reporting and sampling.

---

[17] http://eur-lex.europa.eu/LexUriServ/LexUriServ.do?uri=CELEX:32004R0852R%2801%29:NL:HTML

[18] http://eur-lex.europa.eu/LexUriServ/site/nl/consleg/2004/R/02004R0853-20060101-nl.pdf

[19] http://eur-lex.europa.eu/LexUriServ/site/nl/oj/2004/l_226/l_22620040625nl00830127.pdf

[20] http://wetten.overheid.nl/cgi-bin/deeplink/law1/title=Warenwet

[21] http://eur-lex.europa.eu/LexUriServ/LexUriServ.do?uri=OJ:L:2005:338:0001:01:NL:HTML

2.8.4 Private quality initiatives and systems

Within the poultry meat chain the Action Plan for Salmonella and Campylobacter (www.pve. nl) and the quality system IKB are important private quality initiatives. In both initiatives the Product Boards of Livestock, Meat and Eggs (PVE) play an important role.

The Action Plan Salmonella and Campylobacter

The PVE are aimed at a sustainable social -economic development of their sectors by means of extension, research, subsidies and regulations. The PVE also manage a risk fund to compensate the losses in case of animal diseases. Besides its own activities the PVE co-operate with the government for many tasks in its sectors. The PVE are allowed to announce mandatory regulations, such as the Action Plan Salmonella and Campylobacter but can also make voluntary requirements, for example, in the IKB system (see next paragraph). This plan was introduced in 1997 as a mandatory monitoring system to reduce Salmonella and Campylobacter contamination in all firms of the poultry meat chain, from reproduction to cutting firms and is based on the following principles (Tacken and Van Horne, 2006):
- Taking hygiene measures
- Cleaning and disinfection of buildings
- Inspection of incoming chickens and eggs
- Exchange of inspection results with buyers and suppliers
- Appropriate measures after contamination of poultry meat

During the last three years contamination with Salmonella and Campylobacter of the end products decreased from 20% to 6% (depending on the season with higher contamination levels in the summer). The Action Plan Salmonella and Campylobacter has a strong relation with the private quality system IKB.

IKB

In the poultry meat chain in The Netherlands, 95% of the firms (both primary producers and slaughterhouses and/or cutters) participate in Integraal Keten Beheer, IKB (in English: Integrated Chain Control). IKB encompasses the total Action Plan for the reduction of Salmonella and Campylobacter, but includes additional requirements on traceability, quality and registration. IKB chickens are raised on firms that are regularly inspected by independent organisations on the use of feed, medicine use, hormones, hygiene, but are also inspected on issues such as animal welfare and transport. The PVE are owners of the systems. Within the IKB systems a number of sanctions exist such as warnings, fines, or in the case of repetitive non-compliance exclusion from the system or even closing of the firm. Depending on their performance primary producers are inspected once to four times a year and processors are inspected twice a year. Firms participating in the IKB in The Netherlands can participate in the QS System, by adding an add-on QS module dealing with antibiotics in the feed.

2.9 Regulations for the fruit and vegetable chain

In this section the most important public regulations with regard to food quality and safety for the primary producers and the traders and/or processors in the fruit and vegetable chain are discussed. Next, the inspection situations for these groups of firms are described. Finally, private initiatives for the assurance of quality and safety such as Food Compass and Eurep-GAP are discussed. In comparison with the poultry meat chain, in the fruit and vegetable chain, besides the AID and the VWA some other inspection agencies such as the PD, the KCB and NAKtuinbouw are active, see Box 2.5.

Box 2.5. The PD, KCB and NAKtuinbouw.

The PD (Plantenziektenkundige Dienst, in English:Plant Disease Agency, www.minlnv.nl/pd) is an agency of the Ministry of Agriculture Nature and Food Quality and takes care of the Plant Health in The Netherlands. In the past the PD carried out the phytosanitary inspections. Nowadays the PD supervises the phytosanitary inspections of the KCB and NAKtuinbouw and changes to an expertise centre for Plant Health. Only a few countries, such as Japan, require that the export inspections are carried out by the PD, because of their specific phytosanitary requirements. The KCB (Kwaliteits Controle Bureau, in English: Quality Inspection Agency, www.kcb.nl) was established in 1921 and controls the quality of Dutch export products. NAKtuinbouw (Stichting Nederlandse Algemene Kwaliteitsdienst Tuinbouw, in English: Netherlands Inspection Service for Horticulture, www.naktuinbouw.nl) is responsible for the phytosanitary inspections of propagation materials.

2.9.1 Public regulations

Compared to the poultry meat chain, firms in the fruit and vegetable chain have to comply with a lower number of quality regulations. The food safety risks associated in this chain are lower. Most food safety crises occurred in meat chains. Again, only the most important and well-known regulations for firms are described. Details of other regulations can be found at the provided websites.
- The *Pesticides Law*[22] regulates which and how pesticides are allowed to be used. It also provides specific guidelines for labelling and packaging of pesticides. The VWA inspects this law for the part concerning pesticide residual monitoring in consumer products and the AID inspects for the remaining parts of the law such as the pesticide storage on the firms.
- The *Plant Disease Law*[23] defines the rules for preventing and fighting organisms such as nematodes and diseases such as phytoptera which are harmful for agriculture. In order

---

[22] http://wetten.overheid.nl/cgi-bin/deeplink/law1/title=Bestrijdingsmiddelenwet%201962

[23] http://wetten.overheid.nl/cgi-bin/deeplink/law1/title=Plantenziektenwet

to prevent the emergence and diffusion of harmful organisms, rules are set up for the import and export of propagation materials of plants or plant products. The juridical obligation to inspect fruit and vegetables on phytosanitary requirements lays in the International Permanent Phytosanitary Committee (IPPC) treaty[24] and European phytosanitary guidelines[25]. The Plant Disease Agency and Netherlands Inspection Service for Horticulture (see also Box 2.5).

2.9.2 Inspection situation for growers

In Table 2.10 the inspection situation for growers of fruit and vegetables is displayed. As can be concluded from Table 2.10., the VWA does not inspect growers (unless they have trading activities).

2.9.3 Inspection situation for traders and/or processors

The inspections of traders and/or processors with regard to quality mainly deal with *phytosanitary* and *quality* issues. Moreover, on these firms hygiene aspects are inspected (e.g. the application of HACCP or hygiene guidelines).

Phytosanitary inspections

Phytosanitary requirements deal with the prevention of harmful organisms, so called q-organisms in plants or plant products as demanded in the International Permanent Phytosanitary Committee (IPPC) treaty, the European phytosanitary guidelines[26], and the national legislation such as the Plant Disease Law[27]. An overview of all the regulations that are addressed in the specific phytosanitary requirements is described in the Covenant Plantkeur (www.plantkeur.eu). The essence of these regulations is that for import to and export from the EU fruit and vegetables and flowers and potted plants need a phytosanitary inspection. For trading in the EU propagation materials need a plant passport which ensures that phytosanitary inspections have been carried out (Westerman *et al.*, 2005). The Phytosanitary inspections for import and export of end products are carried out by the KCB, whereas the inspections for propagation material are carried out by NAKtuinbouw. However, member states can apply for reduced phytosanitary checks with regard to import. In The Netherlands, by means of data about phytosanitary inspections of the last three years a product-origin relationships for phytosanitary risks is established which helps the inspection agencies to prioritise their inspections (LNV, 2004a). The inspections are based on the information in the system

---

[24] https://www.ippc.int/IPP/En/default.jsp

[25] http://europa.eu.int/eur-lex/nl/consleg/pdf/2000/nl_2000L0029_do_001.pdf

[26] http://europa.eu.int/eur-lex/nl/consleg/pdf/2000/nl_2000L0029_do_001.pdf

[27] http://wetten.overheid.nl/cgi-bin/deeplink/law1/title=Plantenziektenwet

Table 2.10. The inspection situation for fruit and vegetable growers (adapted from www.inspectieloket.nl).

Inspected issues	Aim of the inspections	Which inspection tasks are conducted	Duration of inspection	% of firms that is inspected	Regulations addressed ***	Inspection agency
User guidelines, user licence, pesticide plan, registration	To promote the food safety and environmental issues by promoting the correct use of pesticides, preventing the use, store, and trade of forbidden pesticides and the prevention of the use of pesticides which are contrary to the user guidelines	Inspection of the inventory, storage and use of pesticides, including the licence	2	10%	1, 2, 6, 9, 10	AID
User guidelines and responsibility with regard to manure production and - use	To promote sustainable production, such as higher quality of water with regard to nitrogen and phosphate concentration	Inspection of administrative obligations with regard to manure production and the requirements for the supply and removal of manure	4	1%	7	AID
Presence of quarantine organisms	Inspection on the presence of quarantine organisms, if they are present, measures could be imposed. When found: Execution of the measures imposed in order to fight the spread of the organisms	Visual inspection and sampling	1 / 1	20% / 100%	3, 4, 5, 8	PD
Minimum requirements of the labour law	To improve compliance in order to enlarge the safety of the employees	Inspection of physical burdens, hazardous stuffs, pesticides, machine safety, etc.	3	3%	11	AI*

Table 2.10. Continued.

* Arbeidsinspectie (in English: Labour Inspection Agency).

Regulations addressed:

1. Decision on integrated crop protection: http://wetten.overheid.nl/cgi-bin/deeplink/law1/title=Besluit%20beginselen%20ge%C3%AFntegreerde%20gewasbescherming

2. Pesticide Law: http://wetten.overheid.nl/cgi-bin/deeplink/law1/title=Bestrijdingsmiddelenwet%201962

3. Regulation on fighting of Phytophera (http://wetten.overheid.nl/cgi-bin/deeplink/law1/title=Besluit%20bestrijding%20aardappelmoeheid%201991) Ralstonia solanacearum (http://wetten.overheid.nl/cgi-bin/deeplink/law1/title=Regeling%20bruin-%20en%20ringrot%202000), Clavibacter michiganensis (http://wetten.overheid.nl/cgi-bin/deeplink/law1/title=Regeling%20bruin-%20en%20ringrot%202000) and Synchytrium endobioticum (http://wetten.overheid.nl/cgi-bin/deeplink/law1/title=Besluit%20bestrijding%20wratziekte%201973).

4. Phytosanitary Directive (RL 2000/29/EU): http://europa.eu.int/eur-lex/nl/consleg/pdf/2000/nl_2000L0029_do_001.pdf

5. International Standard for phytosanitary measures https://www.ippc.int/IPP/En/default.jsp

6. Decision on draining: http://wetten.overheid.nl/cgi-bin/deeplink/law1/title=Lozingenbesluit%20open%20teelt%20en%20veehouderij

7. Manure and fertiliser law: http://wetten.overheid.nl/cgi-bin/deeplink/law1/title=Meststoffenwet

8. Plant Disease Law: http://wetten.overheid.nl/cgi-bin/deeplink/law1/title=Plantenziektenwet

9. Regulation on soil decontamination: (www.minlnv.nl)

10. Pesticide and Biocide Law, in development, will replace the Pesticide Law

11. Law on labour circumstances: http://wetten.overheid.nl/cgi-bin/deeplink/law1/title=Arbeidsomstandighedenwet%201998

CLIENT-import[28] in which firms, the customs and the KCB work together. Importers have to deliver data about the products in this system and the customs and the KCB decide on basis of this information whether or not inspections will be carried out. Only firms that deliver data to CLIENT-import can get reduced checks for phytosanitary inspections. For export the inspection situation of the phytosanitary requirements is based on the requirements of the destination country.

Quality inspections

The European Quality Regulation EG 1148/2001[29] requires that fruit and vegetables are inspected on quality requirements. These requirements are also included in the Dutch *Agricultural Quality Law*[30] and are carried out by the KCB. As far as products need a phytosanitary inspection and a quality inspection, the KCB combines these inspections. Traders of fruit and vegetables can make use of the Regeling Interne Kwalteitscontrole (RIK, in English Regulation Internal Quality Control) in order to get lower inspection frequencies of the KCB. Firms that want to participate in RIK must have (www.kcb.nl):
1. Good results on the quality inspections carried out by the KCB, which means that less than 10% of the batch does not comply with the quality requirements during a period of six months.
2. A quality management system for fruit and vegetables, (the KCB quality code[31]).

For the *imports from outside the EU* of fruit and vegetables the inspection percentages for quality requirements are based on the inspection results of the *previous season* (reduced checks). For firms participating in RIK the constant inspection percentage of 10% of the total number of batches is applied. However, the risk profile for quality and phytosanitary issues are compared with each other and the highest inspection percentage is retained. Also on the *internal European market* the inspection percentages are based on a risk approach which is based on the inspection results of the *previous four weeks*. There are three categories, A, B, and C, in which A is the lowest and C the category with the highest risk profile. The number of inspections of the KCB varies from 2 to 16 times per four weeks depending on the category a firm belongs to. Each four weeks the risk profile of the firms is re-evaluated. Firms participating in the RIK regulation only receive a limited number of inspections per year. For the *export outside the EU* it is required that 100% of batches are inspected. However, for firms participating in RIK this percentage is only *5%*.

---

[28] Controles op Landbouwgoederen bij Import en Export naar een Nieuwe Toekomst (in English: Import and Export Inspections on Agricultural Products to a new future)

[29] http://eur-lex.europa.eu/LexUriServ/site/en/oj/2001/l_156/l_15620010613en00090022.pdf

[30] http://wetten.overheid.nl/cgi-bin/deeplink/law1/title=landbouwkwaliteitswet

[31] www.kcb.nl

Hygiene inspections

According to Regulation EG 852/2004[32], traders and/or processors of fruit and vegetables are obligated to have a HACCP system operational. The Dutch Product Board for Horticulture (PT) has developed a hygiene code for traders and/or processors, which can be downloaded from their site (www.tuinbouw.nl). One of the most important requirements in the hygiene code is that firms have to carry out residual analysis on acute toxic pesticides. The VWA inspects the application and the contents of the hygiene code or the own HACCP system. In order to manage the residual analyses well, Food Compass was introduced (see Section 2.9.4).

2.9.4 Private quality initiatives and systems

In the fruit and vegetable chain Food Compass (www.foodcompass.nl) and the Early Warning and Response System (www.rikilt.wur.nl) are important private initiatives. Furthermore, almost all primary producers comply with the private quality management system Eurep-GAP (discussed in Section 2.6).

Food Compass and the Early Warning and Response System

In order to comply with the *Pesticides Law*[33] and to manage the residual analyses well as required in the hygiene codes, Food Compass was introduced, an independent, Dutch national non-profit making organisation that provides a residue monitoring inspection service to its associate members. Food Compass was established in July 2003 by the Fruit Trade Association Netherlands (Frugi Venta) and the Dutch Product Board for Horticulture (PT). Food Compass is responsible for collection of the data, which are then made accessible for the members. This system is officially approved and checked by the VWA. The outcomes of the tests on the residuals are stored in the database of the Early Warning and Response System (EWRS) managed by the PT. The objective is to reduce the number of exceedings of the norms, to estimate the consequence of exceeding the norms for public health and to prevent exceedings by conducting sector-wide measures. The EWRS system is closely related to the hygiene codes of the PT. The EWRS system consists of three components:
1. A database with residue legislation in The Netherlands and a number of the countries for a number of crops.
2. A database with the results of residual analyses, provided by firms and the VWA
3. Links to other relevant websites for residuals and food safety.

---

[32] http://eur-lex.europa.eu/LexUriServ/LexUriServ.do?uri=CELEX:32004R0852R%2801%29:NL:HTML

[33] http://wetten.overheid.nl/cgi-bin/deeplink/law1/title=Bestrijdingsmiddelenwet%201962

When exceeding of the Maximum Residual Limit (MRL)[34] is detected a number of experts sends a message to the firm concerned and how the firm has to deal with the exceeding. Firms that have applied a hygiene code are obliged to conduct the measures in case of exceeding the norms. Furthermore, the system is able to investigate if the products concerned comply with the legislative demands regarding pesticide residuals in other countries (Bondt et al., 2006).

2.10 Regulations for the flowers and potted plant chain

To a large extent growers and traders in the flower and potted plant chains have to comply with the same regulations as growers and traders in the fruit and vegetable chain, except for the regulations explicitly dealing with food safety and hygiene. Therefore, the public quality regulations and the inspection requirements for growers and traders of flowers and potted plants are not discussed. However, the flower and potted plant chain have their own quality management systems: MPS.

The MPS systems are developed by the foundation Milieu Program Sierteelt, MPS (in English: Environmental Program Floriculture, www.my-mps.com). They developed a certification program aimed at reducing the environmental impact of the floriculture sector and improving the sector's image. The first system MPS with the levels A, B or C was aimed at primary producers. Besides MPS A, B and C other MPS quality systems exist, such as:
• MPS-GAP, which puts more stringent requirements on safety, sustainability and quality compared to MPS A, B, or C.
• MPS Socially Qualified which is based on universal human rights, the codes of conduct of local representative organisations, and International Labour Organization (ILO) agreements.
• MPS Quality responds to the increasing quality requirements set by purchasers, making quality a strategic choice for primary producers.
• MPS Florimark is a certificate that is awarded upon certification for compliance with MPS A, MPS GAP, MPS Quality and MPS Socially Qualified together.

MPS also certifies wholesalers and exporters for systems such as Florimark TraceCert, Florimark Good Trade Practices (GTP), Florimark Trade and MPS TradeCert. These systems are especially dealing with issues such as traceability, ethical trading practices and specific demands of retailers. More information can be found on www.my-mps.com).

---

[34] The setting of MRLs in food is a shared responsibility of the European Union and the Member States. For these pesticide/commodity combinations where no Community MRL exist, the situation is not harmonised and the Member States may set MRLs at national level to protect the health of consumers. MRLs are not maximum toxicological limits. They are based on good agricultural practices and they represent the maximum amount of residue that might be expected on a commodity if good agricultural practices was adhered to during the use of pesticide. Nonethelss, when MRLs are set, care is taken to ensure that maximum levels do not give rise to toxicological concerns (Van der Meulen and Van der Velde, 2004).

2.11 Concluding remarks

In this chapter, the characteristics of the Dutch agri-food sector are described. Although its relative importance for the Dutch economy is declining, it is still one of the most important sectors, especially with regard to the export of Dutch products. The current trend that an increasing amount of products from abroad are being traded and processed in The Netherlands is likely to continue. Good quality is by far the most important requirement for participating in international trade. Furthermore, the three chains included in this study, the poultry meat chain, the fruit and vegetable chain and the flower and potted plant chains have been described in more detail. The huge share of the flower and potted plant chain in Dutch agricultural exports is remarkable.

This chapter has continued by describing the attention that food safety and quality has received from the European and national governments during the last decade. Since the late 1990s, increasingly due to the emerging food safety and quality crises, the European and national governments recognised that a more stringent policy towards food safety and quality was needed. This new approach emphasised quality and safety in the whole agri-food supply chain, so from farm-to-fork. The EU and the Dutch government have aimed to better achieve these goals by setting up food safety authorities such as the EFSA in the EU and the VWA in The Netherlands. The national food safety and quality policies are increasingly derived from European regulations and existing policies are adapted toward European perspectives. However, not only the government has developed measures for the assurance of food safety and food quality, also many private regulations have been developed in the past. These regulations are often initiated by Product Boards, Industry Boards or large retailers and go often beyond legislative quality requirements. Regarding this simultaneous development or public and private quality regulations and the accompanying administrative burdens for firms, the government and firms strive to 'control-on-control', a form of self regulation in which firms get more responsibility for the quality and safety of their products.

To enlarge insight in the inspection situation in the chains, this chapter also gave an overview of the most important regulations and their inspections in the three individual chains. For each chain, the most important legislative quality regulations were described. One should notice that the term 'quality' has a broad scope. In Table 2.11 below is described what is mainly meant by quality in the three chains, based on the most important quality regulations.

This study concludes with another recent study of Bondt *et al.* (2006) that clear overviews of food safety legislation are lacking for agri-food supply chains in The Netherlands, except for primary producers (see www.inspectieloket.nl). This is a problem not only for researchers, but also for firms active in the chain involved in this study. Moreover, it is difficult to judge whether or not legislative demands are included in private systems in an appropriate way. The main explanation is that legislative requirements are general and objective oriented requirements, so called *open norms*. On the contrary, private quality management systems are

Table 2.11. Main topics of the term quality in the selected chains.

Chain	Main quality topic
Poultry meat	Food safety, animal welfare
Fruit and vegetables	Food safety, environmental issues
Flowers and potted plants	Environmental issues, labour issues

means requirements, so called *closed norms*. Especially it would be useful to provide a brief description of how a private quality management system tries to fulfil a legislative demand. This information is likely to be present at the owners of private quality management systems. Therefore, it is recommended that governmental agencies and certifying organisations (of private quality standards) exchange their views on the compliance process.

Despite the difficulties described above, it seems that private regulations can effectively deal with legislative requirements. Both private and public regulations demand firms to work according to HACCP. For example, the hygiene codes of the Dutch Product Board for Horticulture in the fruit and vegetable chain comply with the legislative requirement to have HACCP systems operational, but also assure compliance with quality and hygiene measurements of the private Eurep-GAP system. Furthermore, the control and sanctioning policy of private quality management systems are more transparent than the public requirements. Most of the relevant inspection documents are available on the Internet and the costs of a re-inspection by a certifying organisation can easily be retrieved. Moreover, another very important remark is that the inspection frequency of public regulations is much higher (in some cases even twenty times) than for private quality regulations. Finally, in the three chains investigated, some quality management systems such as IKB, Eurep-GAP are industry wide audited systems which have a participation percentage of 90% or even higher. Participating in such systems is often regarded as a threshold for trading. Although the government presents 'control-on-control' as something new (firms that have quality management systems operational and comply well with them, will get lower inspection frequencies), governmental agencies already work according to a risk based approach for composing the samples of firms that will get an inspection at the moment.

Both the government and the firms are willing to further develop and extend self regulation of quality management in agri-food supply chains. For the government self regulation is advantageous, regulating the market is not a simple job. In developing and maintaining regulations, the government faces many problems which make regulation extremely difficult. Examples are the higher complexity of the reality and the high amount of required knowledge about a problem. For firms self regulation is advantageous too, because they (Baarsma *et al.*, 2003):

1. have a better understanding of practice and own specific knowledge for solving specific problems,
2. are more sensitive to requirements of their buyers than to requirements of the government,
3. and perceive that self regulating entities are less bureaucratic than the governmental regulations.

Besides the high potential advantages of self regulation also disadvantages exist (Balk-Theuws *et al.*, 2004). *Firstly,* the sector itself has to develop and to maintain the system for complying with the conditions set by the government, which can result in higher costs. *Secondly,* some countries (for example, Russia, United States and Japan) may not accept the working practices of self regulation and might consider stopping importing Dutch agriculture products, because they do not trust situations in which tasks of the government are executed by private certification organisations. In addition, due to the different certification systems, the transparency for export countries is lowered.

The literature has introduced a number of success factors for the introduction of self regulation, which should be taken into account (Balk-Theuws *et al.*, 2004). Firms should have:
1. a common interest for complying with the regulations, because the failure of an individual firm can have a negative impact on the total sector,
2. insight in each others behaviour and be able to observe deviations from the regulations, and
3. means to sanction and to reward companies recognising the economic advantages of participating in self regulating systems.

As already argued in the first chapter, these critical success factors can effectively be addressed by collaboration in chains.

Chapter 3. Theoretical review

This chapter discusses theories related to integration and self regulation of quality management systems in agri-food supply chains. It starts with a discussion of the principles of Supply Chain Management in Section 3.1. It describes how business transactions are conducted in supply chains. The content of the term quality (management) is presented in Section 3.2. For the management of quality, the present study focuses on Total Quality Management (TQM) and relates TQM with Supply Chain Management. For choosing appropriate governance forms in the supply chain, Transaction Cost Theory and Contingency Theory provide important theoretical insights; these are presented in respectively Section 3.3 and Section 3.4. Thereafter Section 3.5 sheds light on the concept of self regulation. The chapter ends with concluding remarks and presents the research model of this study in Section 3.6.

3.1 Supply Chain Management

The origin of Supply Chain Management (SCM) has been inspired by many studies like these on the quality revolution, materials management and integrated logistics, the growing interest in industrial markets and networks and influential industry-specific studies (Chen and Paulraj, 2004). At first, the term SCM was used in the logistics literature as an inventory management approach in the 1980s (Van der Vorst, 2000; Trienekens and Beulens, 2001b). During that time suppliers started experimenting with strategic partnerships with their most important suppliers (Tan, 2001). Although the concept encompasses much more, supplier management initiatives and business relations form the core of SCM (Cooper *et al.*, 1997; Chen and Paulraj, 2004). Nowadays, in SCM there is a shift from business integration to collaborative relationships, which means that firms develop increasingly long-term interactive relationships in which partners share information, work together to solve problems, jointly plan for the future and make their success interdependent (Lambert *et al.*, 1998; Spekman *et al.*, 1998). In SCM the ultimate goal is accurate information and a smooth, continual high quality product flow between partners to maximise buyers' satisfaction (Van der Vorst, 2000). Research shows that the most successful supply chains are those which have carefully linked their internal processes to their suppliers and buyers (Frohlich and Westbrook, 2001). For example, retailers purchasing food need reliable partners in order to comply with legislation on liability and food safety and quality. This can only be achieved by joint investments and cooperation (Ziggers and Trienekens, 1999). The advancements and the development of information and communication technology (ICT) have, without a doubt, aided the development and explosive growth of SCM in today's agri-food supply chains (Trienekens and Van der Vorst, 2003). Easing and speeding the exchange of real-time information enables improvement of collaboration throughout the whole supply chain. The intensive and efficient information transfer between firms increases the responsiveness to changes in buyer demands which makes ICT in SCM of paramount importance (Van der Vorst *et al.*, 2002).

Although SCM has received a lot of attention by managers and academics during the last decade, no clear definition of SCM exists. An extensive literature study of SCM definitions is provided by Van der Vorst (2000). Besides, there is a growing number of terms and related buzzwords. Examples are 'demand chain management', 'value chain management', 'network sourcing', 'supply chain pipeline management', 'value stream management', 'support chains', 'integrated purchasing strategy', 'supplier integration', 'buyer-supplier partnerships', 'supply base management', 'strategic alliances', 'supply chain synchronisation', 'network supply chain', etc. (Croom *et al.*, 2000; Chen and Paulraj, 2004; Van der Vorst, 2004). The various definitions are often developed for a particular research or for a specific situation (Claro, 2003). This study will not add to the current confusion of SCM by coming up with its own definition, but will correspond to four commonalities in the definitions of SCM as summarised by Cooper *et al.* (1997):

1. It evolves through several stages of increasing inter-firm integration and coordination; and, in its broadest sense and implementation, it spans the entire chain from primary producer to ultimate end consumer.
2. It potentially involves many independent firms. Thus, managing intra - and inter-firm relationships is of essential importance.
3. It includes the bidirectional flow of products (materials and services) and information, and the associated managerial and operational activities.
4. It seeks to fulfil the goals of providing high buyer value with an appropriate use of resources, and building competitive chain advantages.

Another important notion in the various definitions that exist for SCM is that it takes into account the external environment of a firm (Croom *et al.*, 2000), which could be added as a fifth commonality of SCM definitions.

In Figure 3.1 a typical supply chain structure is presented, placing the firm in the centre of a network of suppliers and buyers. The vertical structure of the supply chain refers to the number of tiers across the supply chain, whereas the horizontal structure refers to the (number of) buyers and suppliers at each level. Obviously the focal firm is not linked to all tiers of the chains directly. In its supply chain the focal firm can have various relationships. With its first tier buyers and suppliers the focal firm has *managed process links* which are critical to the success of the firm. *Monitored process links* are less critical, but still require attention, because these links must be adequately managed by other chain partners. In *not-managed process links* the firm is not actively involved and these are also not critical for the firm. A final category of process links are links between members of the focal firm's chain and non members (e.g. competitors) of the supply chain. All these links affect the performance of the focal firm and its supply chain (Stock and Lambert, 2001). Every link is a relationship. Together they form the network of a supply chain. Thus, how the supply chain is managed depends on how the relationships are organised (Claro, 2003).

Within a supply chain, buyer-supplier relationships might take various forms. Two different schools of thought can be distinguished in literature on business relationship management (Cousins, 2002). The first is the behavioural or humanistic school which compares relationships

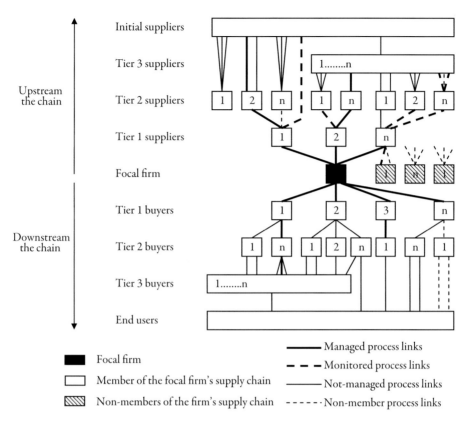

Managed process links
Monitored process links
Not-managed process links
Non-member process links

Focal firm
Member of the focal firm's supply chain
Non-members of the firm's supply chain

Figure 3.1. Supply chain structure (Stock and Lambert, 2001).

between firms as relationships between people like a marriage, based on trust, commitment, mutual understanding and cooperation. The second school takes an economical perspective in which relationships between firms are based on power differences based on differences in size of firms and their economic power in the market. In this study the Transaction Cost Theory is used, which is an example of a behavioural theory. Many authors use Transaction Cost Theory for drawing up a continuum of types of relationships between companies in a supply chain, beginning with market transactions and ending at vertical integration (Cox, 1996; Lambert *et al.*, 1996; Slack *et al.*, 1998; Spekman *et al.*, 1998; Van der Vorst, 2000; Claro, 2003; Verduijn, 2004). Between these two extremes several types of hybrid relationships have been distinguished, see Figure 3.2. The exact names of the types of relationships differ from author to author, but the characteristics of buyer-supplier relationships overlap to a large extent across authors (Verduijn, 2004) .

1. *Spot market relationships* represent market transactions as positioned by Williamson (1985). The fundamental assumption is that trading partners are interchangeable.
2. *Type I (short term focus).* The firms involved recognise each other as partners, and on a limited basis coordinate activities and planning. The partnership has a short term focus.

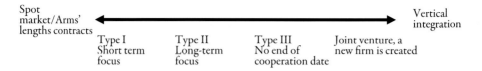

Figure 3.2 Continuum of buyer-supplier relationships (Cox, 1996; Lambert et al., 1996; Slack et al., 1998; Spekman et al., 1998; Van der Vorst, 2000; Claro, 2003; Verduijn, 2004)

3. *Type II (long-term focus).* The firms involved progress beyond coordination of activities to integration of activities, the partnership has long-term horizon.
4. *Type III (no end of date).* The firms share a significant level of operational integration and view each other as extensions of the own firm. No end date for the partnership exists.
5. *Joint ventures* are new created and independent firms separate from the companies forming the alliance. Power in the relationship is based on equivalence.
6. *Vertical integration* or the merger of parties in (part of the) supply chain. In this case all (or part of the) activities from sourcing raw materials to delivering the products to end consumers and supporting activities are coordinated by one firm.

Buyer-supplier relations can take any of the forms discussed above ranging from spot market to vertical integration. Some relationships will be quite rare in the chains studied, such as joint venture and vertical integration; however extreme forms provide a useful analytical baseline from which intermediary forms can be derived (Claro, 2003). Regarding the term 'closely' or 'strongly' integrated chains, the present study focuses on Type II and III relationships.

Since the publication of '*The Wealth of Nations*' by Adam Smith many economists have stressed the ability of free markets (spot markets) to coordinate exchanges leading to the highest utilities for economic actors, the so called '*invisible hand*' (Barney and Hesterly, 1999). However, free markets can only exist if the following three conditions have been fulfilled. *Firstly*, many buyers and suppliers are present in the market which have no individual influence on the market price. *Secondly*, perfect information about the prices and the characteristics of the products exists. *Thirdly*, all economic goods have a price (Baarsma *et al.*, 2003). In practice most markets do not fulfil the conditions for an effective 'invisible hand', resulting in market failure to coordinate economic transactions. Therefore, other governance modes are necessary. In essence, supply chain integration (type II and III) represents a middle ground between markets and vertical integration, capturing advantages while avoiding the risks of both (Ketchen and Giunipero, 2004).

The ultimate goal of SCM is that it leads to significant performance gains (Van der Vorst, 2000). This assumption is also echoed in one of its characteristics, stated before, that it seeks to fulfil the goals of providing high buyer value and building competitive chain advantages (Cooper *et al.*, 1997). An important debate in SCM is the application of appropriate performance measures. A performance measure, or a set of performance measures, is used to determine the efficiency and or effectiveness of an existing system, or to compare alternative systems (Neely *et al.*, 1995). Performance measures are also used to design proposed systems, by determining the values of the desirable levels of performance. Available literature identifies a number of performance measures as important in the evaluation of supply chain effectiveness and efficiency (Beamon, 1998). However, measuring performance of business relationships and especially at the supply chain level is difficult (Gunasekaran *et al.*, 2001; Aramyan *et al.*, 2006). Each partner is likely to adopt their own measurements which might even be conflicting and measures may change over time as the relationship evolves (Lu, 2007). Various studies have developed a number of performance indicators, but their definitions and evaluations are numerous, and only a few definitions are widely accepted.

However, from previous studies it becomes clear that financial performance measures are dominant in empirical studies. Murphy *et al.* (1996) argue that multiple dimensions of performance should be considered where possible, including both financial and non-financial measures. Efficiency and profit are the most commonly used accounting-based performance indicators. In addition, it is important to examine non-financial performance measures such as product quality, buyer satisfaction and market share (Murphy *et al.*, 1996). These indicators of a firm's non-financial effectiveness are what ultimately lead to financial performance. Thus, by taking both dimensions into account, studies in the field of SCM can arrive at accurate estimates of performance (Lu, 2007).

The concept of SCM has also been criticised. Croom *et al.* (2000) and Lamming *et al.* (2000) conclude that most literature about SCM is primarily empirical. Its multidisciplinary origin and evolution have resulted in lack of a robust conceptual framework for the development of SCM theory. As a result the schemes of interpretation of SCM are mostly partial or anecdotic. Moreover, considering the different bodies of literature, such as strategic management, purchasing, logistics, operations management and marketing (Chen and Paulraj, 2004), so many different terms and definitions are not surprising. This study will contribute to the development of SCM theory, by combining it with Total Quality Management. Herewith it supports the research suggestion of Robinson and Malhotra (2005) that quality practices have to change from a firm-based perspective to an inter-organisational supply chain perspective involving suppliers and buyers. Therefore, the combination of SCM and quality management theories can be interpreted as a necessary fundamental extension of the quality management perspective.

3.2 Total Quality Management

Quality has had a lot of attention in management literature e.g. Feigenbaum (1986), Juran (1986), Crosby (1979) and Taguchi (1986). In literature there is a wide spectrum of quality related journals such as Benchmarking: An International Journal, Journal on Quality Management, Total Quality Management and Quality Management Journal. It also receives a lot of attention in more general management journals, such as the Journal on Operations Management and Administrative Science Quarterly. Researchers and practitioners from management, economics, marketing and philosophy in these journals have offered different concepts of quality, which are shown in Table 3.1.

In the transcendent approach quality is equated with 'innate excellence'. It states that although quality is difficult to define, it is absolute and is identifiable through experience. The product-based approach defines quality as the sum or the weighted sum of the scores on desired quality attributes. The user-based approach regards quality as the degree to which it satisfies consumer needs. Manufacturing based definitions of quality equates the concept to compliance with specifications. At least the value-based approach measures quality as the level of compliance with specifications, but at an acceptable cost and price (Forker *et al.*, 1996).

In strategic management literature and research, Reeves and Bednar (1994) found comparable definitions as Forker *et al.* (1996) on the concept of quality, such as 'excellence', 'value', 'compliance-with-specifications', 'and 'meeting and/or exceeding buyers expectations'. In the present study, quality is related to 'compliance-with-specifications', for example, the maximum allowed level of pesticide residuals on a certain crop. The study of Forker *et al.* (1996) also reveals that compliance with specifications is the most important aspect of quality management and has a large impact on the performance of firms. Reeves and Bednar (1994) acknowledge as one of the strengths of this definition that it is relatively easy to measure. Therefore, quality in this study is defined as:

The extent to which firms comply with specifications (or requirements).

Quality has received a lot of attention in business, especially through the introduction of several quality awards. The Malcolm Baldrige National Quality Award was introduced in 1987 by the US government and was established as a statement of national intent to provide quality leadership. Similar awards were created in other industrialised countries, such as the European Quality Management Award and the Deming Application Prize in Japan (Cua *et al.*, 2001). These awards have contributed largely to the conceptual and practical acceptance

Table 3.1. Approaches to defining quality (Forker et al., 1996).

Approach	Definitional variables	Underlying disciplines
Transcendent	Innate excellence	Philosophy
Product based	Quantity of desired product attributes	Management and economics
User based	Satisfaction of consumer preferences	Management and marketing
Manufacturing based	Compliance with specifications	Operations management
Value based	Affordable excellence	Management

of the Total Quality Management (TQM) approach (Samson and Terziovski, 1999; Flynn and Saladin, 2001).

An important practical remark is that in this study food safety (in case of edible products) is regarded as an important part of quality. Quality management systems used in agri-food supply chains include safety (in the case of edible products), often besides a number of other issues, such as health and labour issues. For example, in the Eurep-GAP system quality focuses on the maintenance of consumer confidence in quality and safety, but other important goals are to minimize detrimental environmental impacts of farming operations, optimize the use of inputs and to ensure a responsible approach to worker health and safety (De Bakker *et al.*, 2007). Food quality and food safety management is a complex task due to the inherent characteristics of food and its raw materials (i.e. perishability). In addition, other complicating factors exist such as the large number of linkages in food supply chains and the unpredictable behaviour of people. This human element stresses that food quality management is not only a technological issue. Therefore, Luning *et al.* (2002) promote a techno-managerial approach in which quality problems are considered from an integrated viewpoint of technology and management. A comparable emphasis on technology and management is included in the TQM perspective. In this study it is concluded from an overview of Van der Spiegel (2004) who compared a number of quality management systems in agri-food supply chains, that most quality management systems include TQM principles to ensure proper 'compliance-with-specifications'.

As to the essence of TQM many definitions and frameworks have been proposed in literature, but consensus about the definition does not exist and TQM has been criticised for that (Forza and Filippini, 1998; Luning *et al.*, 2002; Rungtusanatham *et al.*, 2005). A likely explanation could be that TQM theory is far from being fully developed (Ahire *et al.*, 1996) and TQM has become embedded in more and more different firms during the last decades. As a result, it means different things to different people (Sousa and Voss, 2002). However, the TQM philosophy and practice can be reliably distinguished form other strategies and organisational improvements. Therefore, TQM will be the theoretical foundation for the study of quality management in agri-food supply chains.

Several authors have presented sets of empirically validated TQM elements, like Saraph *et al.*, (1989), Ahire *et al.*, (1996), Black and Porter (1996), and Flynn and Saladin (2001). Sousa and Voss (2002) conclude from a comparison between five major instruments for measuring TQM practices, that there is a substantial agreement among the instruments used. They have defined nine common elements for quality management which are: product design, process management, supplier quality management, customer involvement, information and feedback, committed leadership, strategic planning, training and employee involvement, see Figure 3.3. Although TQM makes links to suppliers and customers of a firm, it has so far has specifically focused on quality management within firms.

Figure 3.3. Important elements of TQM (Sousa and Voss, 2002).

Robinson and Malhorta (2005) find that much attention has been paid to Supply Chain Management (SCM) and quality management separately, but that the combination of these concepts has been rare in literature. On the basis of an extended review of articles which lie at the interface of quality management and SCM, they conclude that quality management should further develop from traditional firm centred and product based approaches to inter-organisational supply chain approaches in which all customers and suppliers are involved. Robinson and Malhorta (2005) call this approach Supply Chain Quality Management (SCQM) and define it as:

> The formal coordination and integration of business processes involving all partner organisations in the supply channel to measure, analyse and continually improve products, services, and processes in order to create value and achieve satisfaction of intermediate and final customers in the market place.

Robinson and Malthotra (2005) regard SCQM as a new stage in the evolution of quality management in order to meet the demand for superior products and services (Figure 3.4).

Forza and Fillipini (1998) are among the few authors who have explicitly defined TQM dimensions with customers and suppliers of firms, which are used as a starting point for the application of TQM in a supply chain perspective. For the present study three dimensions could be derived:
1. The TQM approach requires information from suppliers about quality control, i.e. it requires the availability of data concerning the quality of raw materials. This will be regarded as the *monitoring* dimension in this study. Monitoring represents to a large extent the original TQM element process management in a chain perspective (see Figure 3.3).
2. Once it is recognised that characteristics of purchased raw materials may cause quality problems firms will invest in relationships with their suppliers. Often buyers strive for long-term relationships with high performing suppliers. This dimension will be regarded as the *alignment* dimension. Alignment covers to a large extent the original TQM elements supplier and customer involvement.

Programs	-Acceptance sampling -Control charts -Statistical quality control -Inspection	-Zero defects -Problem solving -Quality circles -Statistical process control -Design of experiments	-TQM -ISO 9001 -Baldrige Award -Six sigma	Supply chain management	Supply Chain Quality Management (SCQM)

Years	1920-1960	1960-1980	1980-1990	1990-present	2007 and beyond
Focus	Internal organisation	Internal organisation	-Supply base -organisation -customer expectations	All supply channel members and mostly internal organisation	All supply channel members and mostly internal organisation

Figure 3.4. Evolutionary timeline and focus of supply chain quality management (Robinson and Malhotra, 2005).

3. For a firm it is also essential to maintain close links with its suppliers for quality improvement (for example, feedback on quality performance and customer requirements). Therefore, the *improvement* dimension is included in this study. Improvement covers to a large extent the original TQM element information and feedback.

The selection of the dimensions monitoring, alignment and improvement is also present in the operationalisation of the concept of supplier partnership of Lai and Cheng (2003). These concepts will be used for measuring the integration of quality management in the chain and will further be discussed in the next chapter.

3.3 Transaction Cost Theory

For analysing economic organisation between firms Transaction Cost Theory (TCT) has been arguably the dominant theory (Rindfleisch and Heide, 1997; Masten, 2000; Leiblein, 2003). TCT belongs to the 'New Institutional Economics' paradigm and focuses on governance structures, in which the term 'governance' is broadly defined as 'mode of organisation' (Williamson, 1991). Governance is viewed in terms of particular mechanisms supporting an economic transaction where there is an exchange of property rights. TCT tries to derive the optimal governance mechanism under a certain set of contingencies (Barney and Hesterly, 1999). TCT includes three assumptions that underlie decisions on a given governance mechanism (Rindfleisch and Heide, 1997; Barney and Hesterly, 1999; Barzel, 2000; Masten, 2000; Dorward, 2001; Bijman, 2002; Leiblein, 2003):

1. *Bounded rationality* that refers to the limited capacity of humans to formulate and solve complex problems due to limited availability of information (Simon, 1957). Without cognitive limits, all exchanges could be conducted through planning and it would be possible for actors to write down all requirements in complex contracts.

2. Williamson (1996) recognises that people will behave *opportunistically* in business transactions and people will seek to serve their self interests with guile which makes it difficult to know beforehand who is trustworthy and who not.
3. *Information is asymmetrically distributed.* Thus people only have access to incomplete, imperfect or imbalanced information. TCT assumes that no perfect information exists and exchanges are not costless. TCT considers the efficiency implications of adopting alternative modes of governance in transactions.

The logic of TCT is that collaboration in buyer-supplier relationships strives for the lowest transaction costs (Rindfleisch and Heide, 1997; Bijman, 2002). TCT has conceptualised three general types of governance forms: markets, vertical integration and hybrid or intermediate mechanisms. The characteristics of these governance forms were already discussed in Section 3.1. In TCT any transaction has *ex ante* transaction costs (arising *before* the transaction, such as searching and screening potential exchange agents and bargaining) and *ex post* transaction costs (arise *after* the transaction such as monitoring compliance with contractual terms and enforcing sanctions in the event of non-compliance). According to TCT if transactions costs are low, economic actors will favour market governance. If these costs are high enough to exceed cost advantages of markets, firms will favour contracting or internal organisation (Masten, 2000). In fact, transaction costs are seen as a kind of friction cost for running an economic system (Rindfleisch and Heide, 1997). In the case of strong collaboration, the buyer-supplier relationship is close to vertical integration, whereas lower levels of collaboration implicate spot market forms of collaboration (Claro, 2003). In the original framework to study governance mechanisms Williamson (1985) identifies three main transaction characteristics, namely *transaction specific investments, uncertainty* and *frequency.*[35]

Transaction specific investments refer to the degree to which investments can be used without decrease of value when a relationship is terminated (Williamson, 1975, 1985; Klein-Woolthuis, 1999). Assets with a high amount of specificity are a form of sunk costs (Rindfleisch and Heide, 1997). Examples of transaction specific investments are specificity of location, physical resources and human resources (Williamson, 1991; Benschop, 1997; Poole, 1998); Poole, 1998; Benschop, 1997 Bijman). The idiosyncratic nature of these investments give rise to safeguarding problems and consequently a governance form must be designed to minimise the risk of subsequent opportunistic behaviour (Barney and Hesterly, 1999).

A popular way for firms to achieve safeguarding against opportunism is to integrate with their suppliers, in order to get control over investments and reduce the costs of co-ordination

---

[35] The importance of the dimensions of the transaction differs significantly in the literature. Frequency has received limited attention in TCT literature (Rindfleisch and Heide (1997), because TCT researchers have been largely unsuccessful in confirming the hypothesised effects of frequency on the governance model. Regarding uncertainty the findings are not unambiguous and provide mixed support for the hypothesised effects (Rindfleisch and Heide, 1999). Transaction specific investments have been concerned by many researchers as by far the most important dimension of the transaction and provide the most empirical support for TCT (David and Han, 2004).

and the threat of opportunism (Williamson, 1991; Buvik and Reve, 2001; Humphrey and Schmitz, 2002; Steiner, 2004). Thus, as transaction specific investments increase hybrid (e.g. short and long-term co-operations) and vertical integration mechanisms become the preferred governance mode, see Figure 3.5 (David and Han, 2004).

Although TCT has been used in many studies, there is also much criticism on TCT, see for example Rindfleisch and Heide (1997) and Ghoshal and Moran (1996) for an extensive overview. One of the main drawbacks is that, although Williamson (1991) acknowledges explanations for organisational forms other than transaction costs, such as market power, risk aversion, trust and reputation, TCT is largely neglecting the role of social relationships in economic transactions (Ghoshal and Moran, 1996; Barney and Hesterly, 1999). An example is the fact that previous transactions might influence future transactions (Rindfleisch and Heide, 1997).

Related to this is the vision of Powell (1990) who disagrees with the view of Williamson that hybrid forms of collaboration are mixes between markets and hierarchies. According to Powell this view is '*historically inaccurate, overly static and it detracts the ability to explain many forms of collaboration*'. Powell calls the hybrid forms of collaboration 'networks', in which one party is dependent on resources controlled by another and there are gains to pool the resources. In essence the parties of a network agree to forego the right to pursue their own interests at the expense of others and firms are engaged in reciprocal mutually supportive actions and challenge the view of TCT that individuals are motivated only by self-interest (Powell, 1990).

TCT has especially emphasised the importance of investments in choosing the right governance form, which makes it a very useful theory in this study. In order to assure the quality of products, firms have done transaction specific human and physical investments (Ahire and Dreyfus,

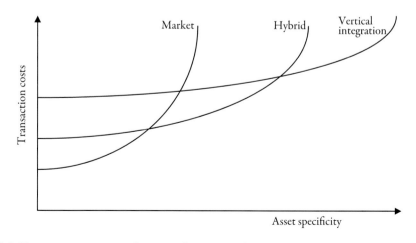

Figure 3.5. Transaction costs as a function of asset specificity (Williamson, 1991).

2000). For example, training of employees to handle the buyer's quality requirements is non salvageable, because a termination of the relationship would ask for learning the specificities of another buyer's quality requirements (Claro, 2003). TCT is widely used in explaining the governance of the relationships of a firm with its buyers and suppliers in agri-food supply chains. However, other theories can also shed light on this. When focusing on the external environment of the agri-food supply chain, Contingency Theory provides insights to what extent firms may integrate their quality management.

3.4 Contingency Theory

Contingency Theory has received considerable attention in organisational theory and strategic management studies (Zajac and Shortell, 1989; Powell, 1992). Almost every theory is contingency based (Ginsberg and Venkatraman, 1985). The starting point of Contingency Theory is that organisations must achieve fit between their external environment and the elements and the structure of their processes (Lawrence and Lorsch, 1967; Boyd and Fulk, 1996). Contingency theorists wish to discover the structural devices and operating methods that ensure long-term survival in different types of business environments (Miller and Friesen, 1978). According to Lawrence and Lorsch (1967) successful firms in more uncertain environments adopt more differentiation and use more sophisticated structures such as integrated systems. For example, differentiation through innovation is typically more necessary in dynamic and uncertain environments (Miller, 1988) than in stable and certain environments. Furthermore, many contingency studies have emphasised a connection between performance and the fit or alignment of the firm within its external environment (Miller, 1988; Venkatraman and Prescott, 1990; Powell, 1992; Doty et al., 1993; Strandholm et al., 2004). Contingency Theory has been used for different topics, such as the examination of the determinants of organisational innovation and the association between technology and structure (Miller and Friesen, 1980). The importance of the external environment has also been evident for the effectiveness of strategic orientation, entrepreneurial orientation and innovativeness (Jansen et al., 2006; Katsikeas et al., 2006).

A lot of hypotheses developed within the Contingency Theory are kind of 'if-then' hypotheses: *if* the environment is dynamic *then* the firm has to differentiate and set up in a sophisticated way (Miller and Friesen, 1978). The dynamic process of adjusting and adapting the firm to the environment is enormously complex, containing myriads of decisions and behavioural adjustments on different organisational levels (Miles et al., 1978). Organisational adaptation is a process, which is best understood by looking at what changes in response to what (Miller and Friesen, 1980). Competitiveness and even the ability to survive can depend on a timely adaptation to environmental changes.

An important debate in the Contingency Theory is the question to what extent the external business environment dominates organisations (environmental determinism) and decision makers and the other way around: the degree to which organisations and decision makers dominate environments (strategic choice). (Clark *et al.*, 1994). An important note for this debate is which kind of environment is aimed at. Many authors make a distinction between the *general* environment and the *task* environment. The general environment includes socio-cultural, economic and governmental pressures, whereas the task environment encompasses product-market and factor-market pressures (Yasai-Ardekani and Nystrom, 1996). The general and the task environment are depicted in Figure 3.6. Regarding the general environment the determinism approach is more likely than strategic choice and for the task environment the other away around. The present study takes both perspectives into account, in other words, it investigates to what extent the integration of quality management in agri-food supply chains is dependent on the quality strategy of the focal firm and to what extent on pressures from the external environment.

The strategy of a firm provides its overall direction by specifying the firm's objectives, developing policies and plans to achieve these objectives and allocating resources to implement these policies and plans (Johnson and Scholes, 1999). This is extremely important, because when the strategy of the firm is strongly focused on supporting and improving quality management in the firm, it could be regarded as a forerunner of effective integration of quality management with its suppliers and buyers. Nowadays, the quality strategy of a firm must not only guide the individual company effects, but also focus on activities to realise supply chain opportunities and achieve collaborative quality advantages (Robinson and Malhotra, 2005). For example, Tan *et al.* (2002) have shown that the firm's internal quality strategy can play a significant role in achieving objectives in buyer-supplier relationships. The Transaction Cost Theory emphasises the development of transaction specific investments as another important predictor for the governance form with regard to quality management in the chain.

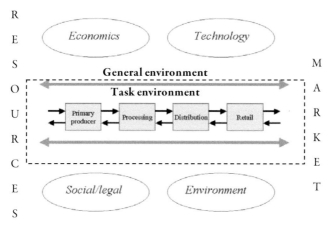

Figure 3.6. The task and the general environment related to the supply chain adapted from Trienekens and Willems (2002).

Although the concept of the external environment is a fundamental concept in management literature, there exists little consensus about its conceptualisation and measurement (Schoonhoven, 1981; Fuentes-Fuentes *et al.*, 2004). Clark *et al.* (1994) has made an overview of the most important dimensions of the environment (Table 3.2) and concludes that two important problems exists for the dimensions of the environment.

1. There is a large overlap and duplication within the dimensions.
2. There may be no end to the dimensionalisation process of the external environment and dimensions could be produced ad infinitum.

Some of the dimensions as summarised by Clark *et al.* (1994) are not applicable for measuring the impact of drivers from the business environment on quality management, for example,

Table 3.2. An overview of environmental dimensions (Clark et al., 1994).

Dimension	Definition	Calibration
Uncertainty	Degree to which probability can be assigned with any degree of confidence to how environmental change will affect the organisation	High/Low
Threat	Degree to which the organisation's goal and priorities are jeopardised by environmental factors	High/Low
Dependency	Degree to which the organisational resources are controlled by environmental forces	Dependent/Independent
Homogeneity	Degree of similarity or difference between the organisation's environmental elements	Homogeneous/Heterogeneous
Dynamism	Degree to which factors identified by decision makers are changing	Dynamic/Static
Rate of change	Frequency and magnitude of turbulence that prevails among environmental factors	High/Low
Routiness	Variability and analysability of environment stimuli confronting the organisation	Routine/Non-routine
Domain consensus	Degree to which the organisation's claim to a specific domain is accepted or disputed by other organisations	Consensus/Dissensus
Turbulence	Extent to which environments are being disturbed by environmental interconnectedness and an increasing rate of interconnection	Many, i.e. turbulent/Non-turbulent
Complexity	Degree to which factors in the organisation's environment are few or great in number and similar or different from another	Simple/Complex
Capacity	Degree of abundance or scarcity of organisational inputs	Rich/Lean

homogeneity, rate of change, routiness, domain consensus, turbulence, complexity and capacity. Regarding the topic of this study to measure the extent to which drivers from the environment have an impact on quality management, the *first* dimension selected is *dependency*. Dependency is the degree to which the firm is controlled or influenced by environmental drivers. Strong dependency means that the behaviour of the firm is altered to a large extent and over a number of decision variables by drivers from the business environment (Jansen *et al.*, 2006). Another assumption is that firms are explicitly paying attention to drivers that critically affect operations and performance, so the dimension *threat* was selected. This *second* dimension threat (or also often called hostility) refers to that aspect of the environmental drivers that poses immediate or potential harm to the firm and its interests. Often high threat levels are associated with more difficulties in controlling the business environment (Miller and Friesen, 1983; Clark *et al.*, 1994). The environment might be hostile for firms whereas other drivers might be favourable; on some drivers the firm is heavily dependent, whereas there is hardly any dependence on others (Yasai-Ardekani and Nystrom, 1996).

Contingency Theory has been criticised by paying too much attention to finding general patterns between the structure of a firm and a given business environment. This approach has led to implementation failure in practice (Ginsberg and Venkatraman, 1985), because it has neglected the characteristics of the firm itself (Miller, 1992). Moreover, the concept environment-structure relationship has a static orientation; in fact it measures the match between the structure and its environment in a single point of time. Up to now, research on environment-structure relationships has not provided methods or tools to cope with the dynamics of a changing environment (Zajac *et al.*, 2000). However, continuous adaptation to a changing environment can move a firm away from its unique competencies (Zajac *et al.*, 2000). Another criticism is that the development of a normative framework for strategic fit is very difficult or impossible. Developing a normative framework requires researchers or managers to prescribe to which, when, in what direction and how many firms should change their structures. Making these kinds of predictions and prescriptions requires firms to know exactly which environmental drivers and organisational structures should be taken into account (Zajac *et al.*, 2000).

Contingency Theory stresses the necessity of maintaining close and consistent linkages between the firm's strategy and the business environment (Venkatraman, 1989). In this study, integration of activities in the supply chain is regarded as a proper solution for dealing with contingencies (e.g. negative media attention or legislation) in their environment, because nowadays no firm is able to handle complex pressures on its own (Gunasekaren *et al.*, 2000; Omta *et al.*, 2002). Thus, in the present study it is assumed that firms which have to deal with strong pressures from their environment will have integrated their processes with buyers and suppliers more tightly than firms which have to deal with lower levels of pressures. If consumers lose confidence in the safety and quality of food, this affects all firms. Therefore, the food industry is actively engaged in setting quality standards, which can be regarded as a form of 'self regulation'. This topic will be discussed in the next section.

3.5 Self regulation

In the past, government carried out most regulative tasks with regard to quality and safety of agri-food products. However traditional *'command-and-control'* regulation by governments is being increasingly replaced in political theory and practice, by alternative, flexible, less state centred forms of regulation (Havinga, 2006). This phenomenon is generally called 'self regulation', which is also known in literature as 'self enforcement', 'self governance' and 'self organisation' (King and Lenox, 2005). Recently, industries in Europe and the United States have argued for a more voluntary approach in which the government sets an overall framework while allowing firms to decide how they organise their processes to achieve the requirements of this framework (Andrews, 1998). Self regulation is not unambiguously defined in literature. Examples of definitions of self regulation are described in Table 3.3. Among these definitions the essence is that a certain group behaves according to a set of regulations which are applied to the members of that group. However, from the definitions above it also becomes clear that the definitions differ with regard to the extent that the government is involved in self regulation, for example, in the definitions of Baarsma (2003) and Havinga (2006). For this study that explicitly focuses on *'control-on-control'* (which will be explained next in this section) the definition of Baarsma *et al.* (2003) fits well. This definition clearly shows that public and private inspections for quality and safety could be complementary to each other and that

Table 3.3. Definitions of self regulation.

Definition	Author
The process in which an organised group regulates the behaviour of its members in order to aim at a certain public interest	Balk Theuws *et al.* (2004)
Societal groups take to a certain extent the responsibility for the formulation, execution and maintenance of rules, if needed within a legal framework	Baarsma *et al.* (2003)
Private actors are regulating the behaviour of their own organisation, its members, or associates without governmental involvement	Havinga (2006)
Voluntary association of firms to control their collective action, as a complement to governmental regulation	King and Lenox (2005)
Process whereby an organised group regulates the behaviour of its members	Gunningham and Sinclair (1998)
Form of setting obligatory norms by a representative board of bodies involved	Geelhoed (1993)
Behaviour regulation which is achieved by means of regulations which are developed by organisations which are not part of the government	Eijlander (1993)
Includes internal control systems that assure product quality, where the company sets, monitors and self-certifies the control parameters	Henson and Caswell (1999)

they could be carried out in a legal framework. This last addition shows that self regulating initiatives of firms may include many requirements that go beyond legislative requirements. Furthermore, many other definitions explicitly assume that there is a certain organisation that supervises its members, which is not the case in agri-food supply chains.

This study takes a continuum on self regulation (see Table 3.4). At the one extreme of 'pure' self regulation firms are completely responsible for the development, execution, inspection and enforcement of requirements, whereas at the other extreme there is a *'command-and-control'* approach which implies that all these tasks are carried out by the government, thus 'pure' governmental regulation. In practice there are mixed forms of 'pure' self regulation and governmental regulation, such as (De Bakker *et al.*, 2007):

1. *Substitution self regulation* in which the government allows (semi) public or private organisations to take care of public interests, but will come up with legislation when these interests are not properly assured.
2. *Legislative conditioned self regulation* in which the government sets the conditional framework and has an important task in the inspection of the end results. However, the (semi) public and private organisations are free to fill in the requirements to a large extent.
3. *Covenants* in which the government makes obligatory agreements, including behavioural regulations with (semi) public and private organisations.

Table 3.4. Regulation forms on a top down/bottom up scale (De Bakker et al., 2007)[1].

Form of regulation	Development, execution, inspection and enforcement of regulations	
'Command-and-control'	Completely by government	No market for certification and inspection
Covenants	Allocation of tasks and expectations are written down in obligatory appointments with (semi) public or private organisations	
Legislative conditioned self regulation ('Control-on-control')	Carried out by (semi) public or private organisations, complying with legislative requirements and are subject of governmental supervision on the results	
'Pure' self regulation	Totally carried out by semi) public or private organisations	Market for certification and inspection

[1] Substitutional self regulation has not been included in this scheme, because it is a form of self regulation that is typically developed for highly skilled professionals, such as advocates and notaries.

This continuum, including the mixed forms of self regulation is depicted in Table 3.4, but it should be noted, however, that in practice many more mixed forms exist. For example, Baarsma *et al.* (2003) have defined 22 self regulating instruments, including certification systems, codes of conduct, and quality management systems.

Most theoretical and empirical literature on self regulation has concentrated on environmental issues, occupational health and safety and privatisation of public utilities. Food quality and food safety are particularly promising fields for self regulation, because of the long history of both public and private forms of regulation, and the current reconfiguration of relationships between the government and firms. Both the government and firms are exploring new ways to assure the quality and safety of food (Havinga, 2006). The Dutch government wrote down the conditions which have to be fulfilled for self regulation in a policy document called *'control-on-control'* (LNV, 2004b).

'Control-on-control' is defined as:

> *Supervisory arrangements whereby the private sector is assigned more responsibility for compliance with statutory regulations; the government operates at a greater distance, but retains the ultimate responsibility. The alignment of the (legal) governmental control activities with the activities of private control organisations is based on agreements with the industry (LNV, 2004b; De Bakker et al., 2007).*

In *'control-on-control'* the key is that firms that comply with legislative requirements will receive lower inspection frequency of governmental inspection agencies. Therefore, *'control-on-control'* can be regarded as legislative conditioned self regulation. The starting point is that participation in these quality management systems is voluntary.

As was already derived from the definition of self regulation the essence is that a certain group behaves according to a set of regulations which are applied to the members of that group. Therefore, the success of self regulation is determined by the level of compliance behaviour of each member. In order to make an estimate about the compliance behaviour of firms, the 'Table of Eleven' can be used (see Table 3.5). The 'Table of Eleven' includes eleven dimensions for compliance grouped according to two dimensions of compliance behaviour, commitment (voluntary compliance) and enforcement (coercive compliance) and is based on the work of Ruimschotel (1994). The 'Table of Eleven' has already been extensively used by many Dutch policy makers and researchers in agri-food chains. For example, the VWA used the 'Table of Eleven' during a pilot study for developing an inspection system in the feed chain (VWA, 2004). Van Amstel-Van Saane (2006) made an application of the 'Table of Eleven' to quality management and evaluated several private quality systems on the dimensions of the 'Table of Eleven'.

Table 3.5. The 'Table of Eleven'.

Commitment	Enforcement
The knowledge and clarity of regulations	The chance that informally discovered offences will be reported to the government
The (im) material (dis) advantages of non-compliance	The perceived chance of discovery of an offence after committing an offence (chance of control)
The degree to which regulations are accepted	The perceived chance of discovery of an offence during an inspection (chance of selection)
The willingness to comply with regulations	The perceived chance of selection for an additional inspection after discovery of an offence (chance of selection)
The chance on discovering and sanctioning by third parties (informal chance getting caught)	The perceived chance of sanction after discovery of an offence (chance of sanction)
	The level and severity of the sanction and additional disadvantages of sanctioning (sanction type).

The central constructs *commitment* and *enforcement* of the 'Table of Eleven' will be further discussed below. These constructs have been central in buyer-supplier relationship literature (Morgan and Hunt, 1994; Hingley, 2005).

Commitment can be defined as an exchange partner's belief that the relationship is worth working on to ensure that it endures indefinitely (Morgan and Hunt, 1994). Definitions and operationalisations of commitment generally encompass three dimensions: *affective, continuance* and *normative* commitment (Allen and Meyer, 1996; Bergman, 2006). Table 3.6 summarises the types of the commitment and describes the motivation of the firm to commit to the relationship. The kind of commitment in the 'Table of Eleven' corresponds with affective commitment which can be defined as the desire to continue a relationship, because of positive affect towards the partner (Kumar *et al.*, 1995). Affective commitment encompasses some elements of benevolence which refers to one firm's belief that the other firms will act in the favour of the party even when there are opportunities for fraud or defect (Kemp, 1999).

According to Morgan and Hunt (1994) a common theme emerging from literature on buyer-supplier relationships shows that firms identify commitment as a key to achieve valuable outcomes. Commitment leads to many advantages in the relationship like working together on investments (Kumar *et al.*, 1995). It goes beyond a simple and positive evaluation of the other firm based on a consideration of the current benefits and costs associated with the relationship.

Table 3.6. Types of commitment in marketing channel literature.

Type of commitment	Based on	Attitude of the firm to relationship
Affective	Identification with, involvement in, and emotional attachment to the other firm	They want to
Continuance	Recognition of the costs associated with ending the relationship with other firm	They have to
Normative	Sense of obligation to the relationship with other firm	They ought to

As already mentioned, enforcement is a form of coercive power and can be defined as the ability of one channel member to influence the decisions and actions of another member by means of punitive actions (Hogarth-Scott and Daripan, 2003). The dependence of buyers on suppliers in the supply chain determines to a great extent their power and use of punitive actions. Integration of quality management in agri-food supply chains often results in a continuous power struggle, making suppliers increasingly dependent on their buyers (Hingley, 2005). Theories dealing with dependence in the supply chain and the use of punitive actions and punitive capabilities as a means to enforce behaviour are the *bilateral deterrence theory*, the *conflict spiral theory* and the *relative power theory*. The overview of these theories is based on Kumar *et al.* (1998). The predictions of these three theories about increasing dependence asymmetry and the use of punitive actions are summarised in Table 3.7.

In case of increasing dependence asymmetry of the partners in the chain the *bilateral deterrence* theory predicts more use of punitive actions of both firms. The less dependent firm knows that it has little to loose when it uses punitive actions. The more dependent firm knows this too and hence renders its fear of retaliation and few restraints on its punitive actions.

Table 3.7. Use of punitive actions in case of increasing dependence asymmetry.

Theory	Use of punitive action in case of increasing dependence asymmetry	
	Less dependent firms	More dependent firm
Bilateral deterrence theory	+	+
Conflict spiral theory	-	-
Relative power theory	+	-

Although the more dependent firms have more to lose than the less dependent firms they expect to be punished by the less dependent firms, regardless of their actions, and therefore, they have a strong motivation to use punitive tactics, pre-emptively to show that it will not passively submit, despite its relative dependence.

According to the *conflict spiral* theory increasing power asymmetry in the chain stabilises relationships, because one firm is clearly dominant over the other. As a more dependent firm faces increasing punitive capabilities, the more dependent firm will avoid provoking actions that will lead to punitive actions of the less dependent firm. The less dependent firm that is aware of this will likely use less punitive actions, because it expects that it will not be harmed by the more dependent firm.

The *relative power* theory offers an alternative explanation of the more dependent firm's behaviour. It asserts that the more dependent firm will be inclined to be as inoffensive and non-threatening as possible so as not to incite the less dependent firm to engage in greater punitive actions, whereas the less dependent firm will be increasingly likely to use punitive actions. The less independent firm has little reason for restrain or fear of retaliation, because it can exit the relationship easily.

Within a buyer-supplier perspective, the integration of quality management can influence both commitment and enforcement. Due to integration firms might believe that there is much to gain by integrating the quality management system in the chain, and that this contributes highly to the overall goals of the firm, increasing the commitment of the firms involved. On the other hand, the integration of quality management might also increase the level of power dependency, offering firms more possibilities for effective sanction. For example, powerful parties in the chain can apply the integration of quality management as a means to obtain more transparency in the production processes of firms which will enable them to sanction deviations from the regulations (Balk-Theuws *et al.*, 2004; Hingley, 2005).

3.6 Concluding remarks

This chapter has described five theories that are of importance for studying quality management and self regulation in agri-food supply chains. Each of these theories offers insight into how firms can carry out quality management and self regulation in their supply chains. At the same time they overlap to some extent and provide rather complementary explanations. So, through a combination of these theories a rich theoretical understanding of the problem being studied can be obtained.

Through the combination of Supply Chain Management and Total Quality Management it is possible to define the most important elements of (integration of) quality management in agri-food supply chains. In addition, Supply Chain Management emphasises the importance of information exchange by ICT which can be regarded as a starting point for a successful integration of supply chain processes. Furthermore, Contingency Theory underlines the importance of external drivers on the way firms organise integration of quality management with their buyers and suppliers. Literature indicates that integration of quality management along the supply chain is the best way for firms to deal with these external drivers. Further, this theory stresses firm strategy to pay attention to the importance of quality management in relationships with suppliers and buyers, while Transaction Cost Theory indicates the effect of transaction specific investments on collaboration in supply chains.

Further, it is argued that the dimensions of compliance behaviour seem to be strongly related to integration of quality management. As a result of integration of quality management in the chain, buyers and suppliers achieve a common interest to comply with quality regulations (commitment). Integration of quality management also results in increased (ICT) transparency in the chain which gives chain partners more possibilities to inspect quality requirements (enforcement).

Finally, it was argued that managing the supply chain as an entity can create better performance. This was recognised as one of the most important characteristics of Supply Chain Management (Cooper and Ellram, 1993; Cooper *et al.*, 1997). Literature on buyer-supplier relationships adds to this that increased performance is likely to be best achieved by means of building mutual interest and commitment and not by means of enforcing objectives (Morgan and Hunt, 1994). Summarising all these thoughts leads to the research model depicted in Figure 3.7. The next chapter of this book will further refine the elements and relationships in the research model. As exemplified by the brief discussion in the first chapter, researches on these topics seem to be quite a promising field of research.

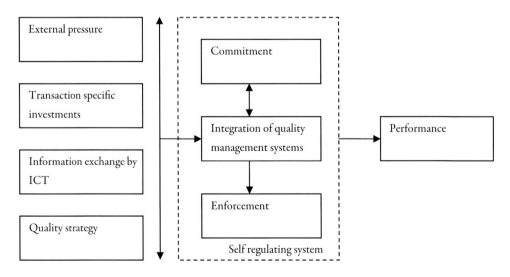

Figure 3.7. Research model.

Chapter 4. Hypotheses

This chapter aims to refine the theoretical content of each concept of the research model presented in the last section of the previous chapter (Figure 3.7). Section 4.1 discusses the factors that are expected to have an impact on integration of quality management. In Section 4.2 the impact of integration of quality management on the dimensions of self regulating behaviour (commitment and enforcement) is described. Section 4.3 offers more insight in the performance measures used and how integration of quality management and the dimensions of self regulating behaviour affect these measures. The chapter ends with concluding remarks in Section 4.4. For measuring concepts dealing with relationships between the focal firm and the supplier and with relationships between the focal firm and the buyer, the same kind of questions were used, making the models comparable (see for the exact formulation of the questions Appendix 4).

4.1 Integration of quality management

In this section the factors are discussed that were expected to have an impact on integration of quality management in agri-food supply chains according to Figure 3.7 in Chapter 3. These factors are *external pressure, transaction specific investments, information exchange by ICT* and the *quality strategy* of the focal firm.

4.1.1 External pressure

Although it may be argued that the business environment exerts the same pressure on all firms in a certain industry, empirical evidence shows that heterogeneity in the business environment exists within one specific industry (Strandholm *et al.*, 2004). Since firms are regarded as open systems, different disciplines have pointed out mechanisms for making a connection between the external environment and the firm's organisation, for instance the work of Miller (1981; 1988; 1992) and Miller and Friesen (1978; 1980; 1983). The connection between the business environment and the firm's organisation that deals with business environmental effects on the strategy of the firm, including its relationships with suppliers and buyers, has received much attention in literature. The central message in these studies is that competitiveness and even the ability to survive depends on a timely adaptation to environmental trends. However, till recently few authors have considered to what extent drivers from the business environment have affected the integration of processes in buyer-supplier relationships, in particular related to quality management (Fuentes-Fuentes *et al.*, 2004). The present study investigates the impact that drivers from the business environment may have on the integration of quality management systems in agri-food supply chains. In Chapter 6 the outcomes from the conjoint analysis show empirical evidence based on 47 expert interviews that the most important drivers for the integration of quality management in agri-food supply chains are *media attention, legislative demands, changing consumer demands* and *societal demands for corporate social responsibility*.

The poultry meat chain especially has received a lot of *media attention* in the last decade, for example, during the dioxin crisis and the Aviaire Influenza crisis (see for example Box 2.1). As a result of these successive crises, *changing consumer demands* have focused on the reinforcement of safety and quality guarantees at all stages in the poultry meat supply chain, from individual farmers to large retailers (Mazé, 2002). Although agri-food products have never met such high food quality standards as nowadays, consumers want to know more about the products they intend to buy than ever before (Rabobank, 2002b). Moreover, the intensive *media attention* of the crises has not only raised consumer awareness of quality and safety issues, but also ethical issues, such as animal welfare and environmental care (Lindgreen and Hingley, 2002; Van Kleef *et al.*, 2006). Society expects firms not only to operate in a sustainable way, but also to demonstrate this to the public. In addition, due to the huge public interest in providing people with high quality and safe food, *legislative demands* regarding safety and quality may provide incentives for firms to engage in effective quality management (Henson and Caswell, 1999) and have been identified by many authors as the most important factor for implementing quality management systems (Downey, 1996). The fruit and vegetable chain shows a comparable trend which sets increasing demands on quality management. For example, in the 1980s and 1990s negative *media attention* harmed the quality image of Dutch products, especially of tomatoes for Germany (Van den Oever, 2005). Many efforts and innovations in products and production processes have been put in place to alter the public quality perception. *Legislative demands* in both the fruit and vegetable chain and the flower and potted plant chain have focused on the reduction of pesticide use. *Societal demands for corporate social responsibility* have put the focus on side-effects of production such as the labour situation, environmental care and energy use. In the flower and potted plant chain, firms have faced societal demands for sustainable production with attention to working practices and ethical trade (Rabobank, 2002b). However, in the flower and potted plant chain, the level of pressures on quality management is expected to be less than in the other two chains, given the fact that flowers and potted plants are non-food products.

Increasingly, the integration of quality management activities throughout the supply chain is seen by authors as the appropriate strategy for dealing with these business pressures (Anderson *et al.*, 1994; Spekman *et al.*, 1999; Morris and Young, 2000; Orriss and Whitehead, 2000; Tuncer, 2001; Grievink *et al.*, 2003). By creating strongly integrated supply chains, ideally the goals of the entire supply chain become the common objective of each firm involved (Lancioni, 2000). In such strongly integrated supply chains, more information will become available to the firms in each stage of the supply chain. For example, tracking and tracing of the whereabouts of the animals and the activities undertaken in the poultry supply chain proved to be essential in preventing the further spread of diseases and in gaining consumer trust. The pressures from the business environment made managers of agri-food firms increasingly aware that incorrect actions at one firm in the supply chain affected the quality assurance and reputation of the complete supply chain and that there was a need for integrated control and intensified cooperation in the whole supply chain (Van der Vorst and Beulens, 2002). Therefore, the following hypothesis is proposed regarding the business pressures and the integration of quality management in agri-food supply chains[36]:

H_1: The higher the pressure from the business environment with regard to product and process quality, the higher the level of integration of quality management systems between the firm and its suppliers (H_1S) and its buyers (H_1B).[37]

4.1.2 Transaction specific investments

Transaction specific investments (TSIs) refer to the transferability of the investments that support a given transaction (Williamson, 1985). Investments with a high specificity represent *sunk costs* that have little value outside a particular exchange relationship. In buyer-supplier relationships, TSIs are usually represented as customisation of products or tailoring of production processes by the supplier on behalf of the buyer (Buvik and Reve, 2001). Many researchers recognise TSIs as the most important dimension that determines the form of governance of a transaction (Rindfleisch and Heide, 1997; Leiblein, 2003; David and Han, 2004). Williamson (1989) has distinguished five types of TSIs:

1. *Site;* the supplier and buyer are located close to each other in order to reduce transportation and inventory costs.
2. *Physical*; investments that tailor processes to particular exchange partners such as customised machinery, tools and dies.
3. *Human;* skills, knowledge and experiences of personnel that are specific to the requirements of another firm.
4. *Dedicated*; generic investments that exceed the level of investments needed if the firms were not engaged in a specific relationship.
5. *Brand name*; worthless if the products to which the brand name is attached are no longer available.

Research has primarily centred on the human and physical dimensions of TSIs (Claro, 2003; Grover and Malhotra, 2003). The production of quality products is necessarily dependent on the use of raw materials, so it is essential that these raw materials meet buyer's quality specifications (Shin *et al.*, 2000; Kaynak, 2003). Buyers who are concerned with the procurement of high quality raw materials request that the suppliers make quality transaction specific investments

---

[36] These drivers (*media attention, legislative demands, changing consumer demands and societal demands for corporate social responsiblility*) were combined into one index, representing the impact from the business environment on the focal firm. The next chapter, study design, presents the way in which this index is calculated. However the hypothesis is also valid for the individual drivers such as media attention, In that case the hypothese would be: The higher the level of media attention with regard to product and process quality, the higher the level of integration of quality management between the focal firm and its suppliers (H_1S) and its buyers (H_1B).

[37] The S and the B refer respectively to the supplier and buyer model.

and adaptations that will improve quality management. For example, firms might have to invest in specific physical assets (for example, cooling or packaging equipment) to comply with quality requirements, or have to invest also in specific human assets in order to increase the level of knowledge and experience specific to assuring the quality of products and processes. The level of TSIs may also vary. For a specific packaging machine that a supplier must purchase to sell products to one particular buyer the purchase is sunk to a large extent and the salvage value would be low when the relationship is terminated. Likewise a supplier's alignment of his quality management system to handle the buyer's specific quality requirements is non-salvageable, because when the relationship ends, it will be necessary to align the quality systems with another buyer. Yet if the investment is to some extent transferable, the degree of specificity becomes lower (Claro, 2003). Moreover, the level of TSIs can also vary *in* chains. Growers of potted plants for instance usually have made large TSIs, because there is a small number of buyers for their products, whereas growers of flowers sell their products through the auction clock, which means that there are less integrated buyer-supplier relationships. Also *between* chains differences may exist. In the poultry sector there is a very small number of slaughterhouses with largely comparable quality requirements, whereas in the fruit and vegetable chains, TSIs may be lower, because there is a large variety of buyers with significantly different quality requirements.

TSIs are important mechanisms for establishing stable buyer-supplier relationships. The deliberate creation of specific investments for the purpose of making it difficult for a partner to end the relationship will also provide sufficient reason for the buyer and supplier to continue to work closely together, especially when firms are highly dependent on each other (Williamson, 1985). The self imposed exit barrier provides incentives for the supplier to live up to its promises, suggesting that TSIs act as a safeguard against opportunistic behaviour. Additionally, TSIs provide a powerful signal to the buyer and are more than hollow promises. Observing the other party's investments causes a buyer member to be more confident in the supplier's commitment to the relationship, because the other firm will sustain economic consequences if the relationship ends (Heide and John, 1988; Anderson and Weitz, 1992).

Initially, the central proposition in Transaction Cost Theory (TCT) was that high levels of TSIs would affect buyer-supplier relationships negatively by fostering dependence and other governance hazards such as opportunism, which can work negatively for the investor, in this case the supplier. However, TSIs do not only create dependence of the investor in the relationships, but also for the other firms involved (Heide and John, 1990). For example, when the buyer requests the supplier to invest in specific assets, both parties will involve in cooperative efforts to make optimal use of resources. In such situations, buyers want to exercise some control and influence over the production and logistics of the supplier in order to have an effective and optimal use of investments. Therefore, bilateral dependence will increase and the need for coordinated adaptation will occur (Williamson, 1991; Buvik and Halskau, 2001; David and Han, 2004). The important thing is that deployment of TSIs by the supplier is expected to provide cost reductions and added value which are beneficial for both the buyer and the supplier. For example, Zaheer and Venkatraman (1995) found a very strong relationship between specific investments and joint planning. Therefore, if partners in the chain have invested to comply with specific quality requirements, transactions should be carried out in close collaboration and in extreme cases even vertically integrated which is in fact one of the central hypothetical predictions of TCT (Williamson, 1975). This leads to the following hypothesis:

H_2: *The higher the level of transaction specific investments of the suppliers cq the firm, the higher the level of integration of quality management systems between the firm and its suppliers (H_2S) or its buyers (H_2B).*

4.1.3 Information exchange by ICT

Information exchange by ICT is defined as the use of information and communication technology (ICT) to connect chain partners in both directions (upstream and downstream). ICT has been recognised in Supply Chain Management as one of the driving forces in creating effective buyer-supplier relationships (Cooper *et al.*, 1997; Lambert and Cooper, 2000; Swartz, 2000; Hill and Scudder, 2002; Chen and Paulraj, 2004). Information exchange by ICT supports supply chain co-ordination, particularly when technologies are used to span the traditional firm boundaries by eliminating the barriers that exist between independent actors in the supply chain (Kumar and Van Dissel, 1996; Croom *et al.*, 2000; Cramer, 2004; Handfield and Nichols, 2004; Matopoulos *et al.*, 2004; Van der Zee, 2004). Higher levels of integration in supply chains are stimulated by more standardised and automated interfaces and flows of data (Hill and Scudder, 2002). Leek *et al.* (2003) have shown that the range and depth of the exchanged information are also altered by information exchange by means of ICT in the supply chain. Regarding quality management systems, information exchange by ICT plays an important role, because it facilitates the exchange of large quantities of quality data between suppliers and buyers (Trienekens and Van der Vorst, 2003). Many quality management systems include requirements for receiving in-process and final inspection data of every stage of the supply chain (Petersen *et al.*, 2002). The systematic gathering and sharing of these large amounts of quality data are extremely important for firms, because of the demands of both governments and retailers to guarantee quality, composition and origin of their products (Trienekens and Beulens, 2001a). Moreover, information exchange by ICT enables the partners to monitor the costs of the different chain partners (Salin, 2000). Taking these features of integrated information exchange into account, the following hypothesis is proposed:

H_3: *The higher the level of information exchange in the chain by means of ICT, the higher the level of integration of quality management systems between the firm and its suppliers (H_3S) or its buyers (H_3B).*

4.1.4 Quality strategy

Strategy can be defined as the overall direction of the firm, by specifying the firm's objectives, developing policies, plans to achieve these objectives and allocating resources to implement policies and plans (Johnson and Scholes, 1999). Strategy explores the firm's strategic advantages in relation to its business environment, while at the same time a firm's strategy must be executable within the firm itself (Klassen and Angell, 1998; Fortuin, 2006). Regarding the strong pressures in the business environment for adequate quality management, many studies have recognised the quality strategy of a firm as very important for the successful implementation of quality management systems (Saraph *et al.*, 1989; Flynn *et al.*, 1995; Ahire *et al.*, 1996; Black and Porter, 1996; Flynn and Saladin, 2001; Kaynak, 2003). Successful implementation of quality management systems in many cases requires the effective change of a firm's culture. Quality culture is almost impossible to change without a concentrated management effort aimed at continuous improvement, open communication and cooperation. For example, involving employees requires the communication of a clear strategy to improve quality enhanced by instituting quality-based compensation and other incentives (Kaynak, 2003). The role of the strategy of the focal firm becomes even more important when quality management is embedded in buyer-supplier relationships (Ellram, 1995; Forza and Filippini, 1998; Krause, 1999; Chen and Paulraj, 2004; Robinson and Malhotra, 2005). In that case management of a firm must not only guide and direct the firm's own quality efforts, but also encourage quality measures among buyers and suppliers (Chen and Paulraj, 2004). In these relationships the importance of shared goals and strategies, the implementation of joint quality initiatives and an integrative focus on communication should be stressed in order to realise successful buyer-supplier relationships (Chen and Paulraj, 2004). For example, Humphreys *et al.* (2004) found that by introducing joint quality standards due to the integration of quality management, the exchange partners will establish a joint platform for future cooperation, reducing transaction costs (Bredahl and Zaibet, 1995; Holleran *et al.*, 1999; Loader and Hobbs, 1999; Reardon *et al.*, 2001). Consequently the following hypothesis is proposed:

H_4: *The more the firm pursues its quality strategy, the higher the level of integration of quality management systems between the firm and its suppliers (H_4S) and its buyers (H_4B).*

4.2 Self regulating behaviour: commitment and enforcement

Integration of quality management is expected to influence the two most important dimensions of self regulation (Balk-Theuws *et al.*, 2004). The essence of integrated quality management systems in agri-food supply chains is to create collaboration in which partners extensively share information and work together to solve problems (Krause and Ellram, 1997; Spekman *et al.*, 1998; Shin *et al.*, 2000). Due to integration efforts and delivering quality products to buyers, buyers and suppliers achieve a common interest to comply with quality regulations (commitment). Furthermore, due to the increased transparency in the chain non-compliance is easier to detect, which gives chain partners more possibilities for inspecting their quality regulations (enforcement). This section will further elaborate on the relationship between integration of quality management and the dimensions of self regulating behaviour.

The relationship between commitment and integration of quality management behaves like a 'spiral' that, based on positive or negative experiences, can go up or go down. On the *one* hand, for firms that want to integrate their quality management the selection of committed suppliers may be very critical and may be the most important indicator for delivering superior raw materials (Shin *et al.*, 2000). Suppliers that have the same attitudes with regard to quality will be highly valued and buyers are very willing to integrate and maintain relationships with such partners (Morgan and Hunt, 1994). The explanation is that commitment goes beyond a simple and positive evaluation of the other firm based on a consideration of the current benefits and costs associated with the relationship (Kumar *et al.*, 1995). As a result firms have a strong motivation to build, maintain, strengthen and integrate the relationships.

On the *other* hand, a major focus of integration of quality management systems in the chain is the creation of strong, collaborative buyer-supplier relationships with joint objectives. This is realised by conducting activities such as the analysis of information about buyer needs, working with suppliers to improve the performance of (second-tier) suppliers, provision of feedback about the quality performance and the measurement of buyer satisfaction (Ahire and Golhar, 1996; Samson and Terziovski, 1999). In closely integrated supply chains buyers regard their suppliers as extensions of their own firm and take a long-term orientation toward them and are willing to make short-term sacrifices to realise long-term benefits from the relationship. The ultimate goal of integration of quality management in the chain is improved performance through the better use of each other's quality management processes creating a seamlessly coordinated supply chain quality management system (Robinson and Malhotra, 2005). In such systems, supply chain sub-optimisation is prevented which occurs if each firm in a supply chain attempts to optimise its own results rather than to integrate its goals and activities with other firms in order to optimise the quality performance of the total chain (Cooper *et al.*, 1997). Literature stated that due to the creation of a common goal and prevention of sub-optimisation, interests of all firms in the agri-food supply chain converge and commitment to quality regulations will emerge or increase, because the firms receive valued contributions from each other. Taking together these thoughts leads to the following hypothesis:

H_5: *The higher the level of commitment of the suppliers to the quality requirements of the buyers, the higher the level of integration of quality management systems between the suppliers and buyers (H_{5a}). And vice versa: The higher the level of integration of quality management systems between the suppliers and buyers, the higher the level of commitment of the suppliers to quality requirements of the buyers ($H_{5a}B$).*

Besides commitment, the integration of quality management is expected to also have an impact based on the other dimension of compliance behaviour, enforcement. As already stated, integration of quality management implies an increased exchange of quality related data, such as the outcomes of quality tests and - inspections. Traditionally, incoming goods inspection and supplier assessment schedules have been the most important mechanisms in monitoring compliance with quality requirements. These systems are used to assess the reliability and overall quality of suppliers in which statistical tests and inspections remain the primary methods employed. Since the recognition that food quality is basically the task of the whole agri-food supply chain, these tests have been supplemented and incorporated in certified quality management systems, which ensure uniform inspections throughout the supply chain (Jahn *et al.*, 2004). In this way, the integration of quality management in agri-food supply chains increases transparency, which is needed to guarantee the safety and quality of food and to control incidents when they occur. Therefore, some authors such as Grievink *et al.* (2003) and Hingley (2005), following the relative power theory (see Section 3.5), have concluded that although the word collaboration is often used with buyer-supplier relationships, the most important objective for buyers is the enforcement of their quality requirements to their suppliers. Spekman *et al.* (1998) also stated that the development of integrated quality management systems in buyer-supplier relationships is difficult and requires a lot of efforts. Therefore, firms that have spent a lot of time and money integrating quality management might increase the frequency of inspections and the severity of sanctions when their suppliers do not comply with their quality management systems. According to Hingley (2005) suppliers in agri-food supply chains broadly accept this as long as this method of doing business has the lowest transaction costs. These thoughts lead to the following hypothesis:

H_6: *The higher the level of integration of quality management systems between the firm and its suppliers or buyers the higher the possibilities for enforcement of the quality requirements by the firm (H_6S) cq the buyers (H_6B).*

Each firm involved in integrated chains has to find the right balance between the use of commitment and enforcement in order to retain well-performing and workable relationships with their buyers and suppliers. According to Morgan and Hunt (1994) a common theme emerging from literature on buyer-supplier relationships is that firms identify commitment and not enforcement as a key to achieving valuable outcomes for themselves. Commitment results in many advantages in the relationship like working at preserving relationship investments and co-operating with suppliers who resist attractive short-term alternatives in favour of expected long-term benefits. Firms review potentially high risk actions prudently, because they do not expect their committed suppliers to behave opportunistically. When commitment is established, firms learn that coordinated joint efforts will lead to outcomes that exceed the outcomes a firm would achieve alone (Anderson and Narus, 1990). However, a certain level of enforcement in buyer-supplier relationships is needed because even if the vast majority of firms does the right thing, there is always a chance that an individual firm will cause serious harm. Agri-food supply chains have a long history of enforcement of quality requirements by manufacturers, trade associations, and corporate organisations, in which the integration of quality management systems in agri-food supply chains has played an important role. However, too strong a focus on enforcement to gain compliance will result in conflict of a dysfunctional kind which will destroy the willingness to co-operate, inhibit long-term success and ultimately lower performance (Morgan and Hunt, 1994; Lazzarini *et al.*, 2004). Therefore, the enforcement based instrument should be used with caution given the uncertainty involved in mobilising it successfully and the risk that its use is perceived to be unjust or unreasonable (Grabosky and Gunningham, 1998). Therefore, this study expects that commitment more than enforcement will show a mediating[38] effect between the integration of quality management and performance.

4.3 Performance

Performance can be defined as the extent to which goals are achieved (Claro, 2003). Many studies have emphasised the positive effect of integration of quality management in buyer-supplier relationships on performance in terms of buyer satisfaction and financial performance. See for an overview for example, Aramyan (2006), Beamon (1999), Claro (2003), Gunasekaran (2001) and Shephard and Günther (2006). Moreover, these two performance indicators have also been frequently used for measuring the effect of quality management (Sousa and Voss, 2002; Kaynak, 2003).

Satisfying buyers is also one of the main characteristics of Supply Chain Management as defined by Cooper *et al.* (1997). For example, Forza and Filippini (1998) found that integration of quality management in the relationship with buyers had a positive relationship with buyer satisfaction. Rungusanatham (1998) also found a positive relationship between quality improvement activities and buyer satisfaction. In addition, Flynn and Saladin (2001) have shown that many aspects of the US-based Baldrige Quality framework show a positive relationship with buyer satisfaction. Furthermore, buyer satisfaction is chosen as a criterion in this study, because the objective of supply chain quality management is '*to create value and achieve satisfaction of intermediate and final buyers in the marketplace*' (Robinson and Malhotra, 2005).

Integration of quality management in the supply chain can influence *financial performance* (Sousa and Voss, 2002). *Firstly*, quality improvements in the chain may result in higher prices and in more satisfied consumers. If the costs of these quality adaptations are lower than the savings, profitability gains can be achieved. *Secondly*, besides the fact that improvement of internal business processes reduces costs through higher efficiencies in the chain, quality management also focuses on the organisational efforts to satisfy buyers, which leads to

---

[38] Mediating effect is caused by the presence of a mediator variable. A mediator variable explains the relationship between two other variables. The general test for mediation is to examine the relation between the independent and the dependent variable, the relation between the independent and the mediator variable, and the relation between the mediator variable and the dependent variable. All of these relationships should be significant. The relationships between the independent and the dependent variable should be reduced (to zero in the case of total mediation) after controlling the relation between the independent variable and mediator and the mediator and the dependent variable (Baron and Kenny, 1986).

increased profitability by gaining market advantage. A financial indicator such as revenue growth was chosen because *'it is the ultimate aim of any business to make money'* (Demirbag *et al.*, 2006). An important notion for the relationship between financial performance and integration of quality management is that financial performance is contingent on many other factors, such as the nature of the markets, the market size and competition (Price and Chen, 1993; Karmarkar and Pitbladdo, 1997; Zhao *et al.*, 2004). However, most studies have shown that quality management has a positive influence on financial performance (Powell, 1995; Grandzol and Gershon, 1997; Easton and S.L, 1998; Kaynak, 2003; Fuentes-Fuentes *et al.*, 2004) also for small and medium sized firms (Demirbag *et al.*, 2006). Few other studies did not find it, for example, Terziovski *et al.* (1997) and Mohrman (1995). Taking all these thoughts about the effects of integration of quality management on performance together leads to the formulation of the following hypothesis:

H₇: The higher the level of integration of quality management systems between the firm and its suppliers (H_7S) or its buyers (H_7B), the higher the buyer satisfaction will be and ultimately the financial performance of the focal firm.

4.4 Concluding remarks

This chapter elaborated on the study's general research model and its expected relationships (Figure 3.7). It presented the most relevant dimensions of the business environment and other factors which have an impact on the integration of quality management. Furthermore, it discussed the impact of integrated quality management on the dimensions of compliance behaviour; commitment and enforcement. For measuring the performance of integrated quality management and self regulation in supply chains, buyer satisfaction and a financial indicator were selected. These performance indicators were expected to be positively influenced by the integration of quality management, especially when integration of quality management results in increased commitment of exchange partners.

The research model was applied to the supplier and buyer side of the focal firm and hypotheses were formulated for both the supplier and buyer model The hypotheses are summarised in Table 4.1 and will be quantitatively tested in Chapter 7.

It is important to recognise that the development of relationships in agri-food supply chains is not always a sequential process (Claro, 2003). However, by formulating hypotheses some sequence and causality basis is introduced. One should realise that the present study positively or negatively cut off the 'spiral' in a buyer-supplier relationship and based on former positive or negative experiences it can go up or go down. For example, due to the on-going integration of quality management, suppliers might decide to invest even more in quality transaction specific assets for a certain buyer.

Table 4.1. The hypotheses for the supplier and buyer model. S refers to the supplier model, B to the buyer model.

Hypotheses

1. The higher the pressure from the business environment with regard to product and process quality, the higher the level of integration of quality management systems between the firm and its suppliers (H_1S) and its buyers (H_1B).

2. The higher the level of transaction specific investments of the suppliers cq the firm, the higher the level of integration of quality management systems between the firm and its suppliers (H_2S) or its buyers (H_2B).

3. The higher the level of information exchange in the chain by means of ICT, the higher the level of integration of quality management systems between the firm and its suppliers (H_3S) or its buyers (H_3B).

4. The more the firm pursues its quality strategy, the higher the level of integration of quality management systems between the firm and its suppliers (H_4S) and its buyers (H_4B).

5. The higher the level of commitment of the suppliers to the quality requirements of the buyers, the higher the level of integration of quality management systems between the suppliers and buyers (H_{5b}). And vice versa: The higher the level of integration of quality management systems between the suppliers and buyers, the higher the level of commitment of the suppliers to quality requirements of the buyers (H_{5b}).

6. The higher the level of integration of quality management systems between the firm and its suppliers or buyers the higher the possibilities for enforcement of the quality requirements by the firm (H_6S) cq the buyers (H_6B).

7. The higher the level of integration of quality management systems between the firm and its suppliers (H_7S) or its buyers (H_7B), the higher the buyer satisfaction will be and ultimately the financial performance of the focal firm.

Chapter 5. Study design

In the present study, a 'mixed methodology' in three constructive phases was used to conduct the empirical research (see Figure 5.1). 'Mixed methodology' designs incorporate techniques from both the quantitative and qualitative research traditions, yet combine them in unique ways to answer research questions that can not be answered in another way. It is expected that 'mixed methodologies' will be the dominant methodological tools in the social and behavioural sciences during the 21st century (Tashakkori and Teddlie, 2003a), because it has been proved to increase the 'goodness' of the answers (Currall and Towler, 2003). 'Mixed methodology' is sometimes referred to as *triangulation* (Jick, 1979; Verschuren and Doorewaard, 1999; Tashakkori and Teddlie, 2003b). The assumption of 'mixed methodology' is that weaknesses of one method can be compensated by the strengths of other methods offering a greater potential for consistent theory building (Wacker, 1998). 'Mixed methodology' can answer research questions that single methodologies cannot, provides better (stronger) interferences and provides the opportunity for presenting a greater diversity of divergent views (Morse, 2003; Teddlie and Tashakkori, 2003). More specifically, in this study, the survey overcomes limited generalisability or external validation of in-depth interviews, while in-depth interviews provide information on how firms work in their single, natural setting, things that are hard to include in surveys.

The *first* phase has an explorative nature and starts with the identification, description and ranking of drivers acting on quality management in agri-food supply chains. It aims to answer questions such as *What are important drivers to improve quality and safety management and how large are their individual impacts? Do the impacts of these drivers on quality and safety management differ for the three chains?* and *How do these drivers influence quality management in the selected chains?* In order to answer these questions a conjoint analysis was conducted

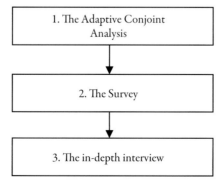

Figure 5.1. The three phases in the present study.

with experts from business and academia. Besides the ranking of the drivers emphasis is on hypotheses formulation based on the experts' motivation for their rankings of the drivers. The design of the conjoint analysis combined with additional interviews (for hypothesis formulation) is described in Section 5.1.

The *second* and most important phase, the survey, was used for quantitative research. The methodology allows researchers to gain an overall picture of a phenomenon. However, a major drawback of this research strategy is that it offers little contextual information. The primary goal of the survey was to test the hypotheses defined in Chapter 4. The survey answers questions such as: *How, and to what extent do factors influence the integration of quality management? How does integration of quality management influence the dimensions of compliance behaviour? and How do integration and self regulation of quality management influence performance of firms in agri-food supply chains?* In quantitative research the problems of reliability and generalisability are easier to address than in a case study. A survey is characterised by large numbers of research units, labour intensive data generation, breadth rather than depth and quantitative data analysis (Verschuren and Doorewaard, 1999). In the context of a survey, a topic (for example commitment) is labelled, defined and operationalised so that it can be measured using numbers and scales. The question of validity concerns the legitimacy of the translation steps, described above, that have been made. Careful definition of indicators can ensure the measurability of concepts. To ensure reliability the survey was extensively pre-tested among a number of potential respondents in each chain. The design of the survey, including the operationalisation of the factors, is described in Section 5.2.

In the *third* phase, the findings from the survey were verified using in-depth interviews with experts in the chains. The objective of this phase was to gain feedback on the results and practical insights in how predicted relationships found in phase two actually happen in practice, counteracting the major drawback of a survey offering little contextual insights. Statistical findings from the survey are combined with statements from these case interviews and can be regarded as 'theory confirmative in-depth interviews'. Another objective of this phase was to come up with answers to the question: *How can 'best practices' quality management systems including self regulation be designed?* By answering this question, the present study provides recommendations for both managers and policy makers. The design of the in-depth interviews is described in Section 5.3.

5.1 The conjoint analysis

This section describes the methodology employed to set up the adaptive conjoint analysis, the first phase in this research. Since few studies have been done to explore the impact of the business environment on quality management (Fuentes-Fuentes *et al.*, 2004), the study started with the identification, description and ranking of drivers acting on agri-food supply chains. From an extensive literature research and explorative interviews with experts participating in the EU concerted action 'Global Food Network'[39], the main drivers that are supposed to impact on quality management were selected. The main objective of this phase is to refine thoughts about drivers for quality management and to quantify their impact on quality management in agri-food supply chains. It has to answer the following research questions:

1. What are important drivers for quality and safety management in agri-food supply chains and how large are their individual impacts on quality management?
2. Do impacts of these drivers on quality management differ for the three chains?
3. How do these drivers influence quality management in the three chains?

5.1.1 What is a conjoint analysis?

The objective of conjoint analysis is to determine what combination of a limited number of attributes of, for example, a product or service, has the largest impact on a respondent. For example, a car may have attributes such as colour, size, brand, and so on. Each attribute can then be broken down into a number of levels. Levels for colour format may be red, blue, black, white, etc. Respondents would be shown a set of pictures of cars created from a combination of levels from all or some of the constituent attributes and asked to rank or rate the cars according to their buying preference. Each car is composed of a unique combination of attributes and each car is similar enough so that consumers will see them as close substitutes, but dissimilar enough so that respondents can clearly determine a preference. While traditional (so called compositional) analyses evaluate the different attributes separately[40], conjoint analysis evaluates the multiple attributes and attribute levels 'jointly' (hence the term conjoint analysis) to, identify the car that contains the most preferred combination of attributes for the respondent. The ranking of the cars is then split into judgements of the individual attributes. The main advantage of compositional analysis is the relatively simple method of data collection and the limited cognitive demand required of the respondents. Conjoint analysis better represents real life conditions, and thus provides a more realistic picture (Agarwal and Green, 1991; Churchill, 1999; Van der Haar et al., 2001).

Although conjoint analysis has extensively been used in marketing applications, it can also be used for research in agri-food supply chains (Hobbs, 1996; Skytte and Blunch, 2001; Meuwissen et al., 2004). In this study, conjoint analysis is specifically used to allow firms to

---

[39] The EU-concerted action 'Global Food Network' (www.globalfoodnetwork.org) consisted of countries from the European Union, Latin America and Africa. It is an international knowledge network on international food supply chains and networks in which public and private organisations (universities, research institutions, governmental organisations and agri-food firms) participate (Trienekens et al., 2005). Food safety and quality are the main focus areas. Each country has made a report in which national agri-food industry trends, bottlenecks and opportunities regarding food safety and quality for international meat, fish and fruit vegetables chains have been specified.

[40] An example of a frequent used compositional approach is the Analytic Hierarchy Process (AHP) proposed by Saaty (1977). Within AHP a compositional approach is used, where a multivariable problem is first structured into a hierarchy of interrelated elements, and then a pairwise comparison of variables in terms of their dominance is elicited (Muyle, 1998). In the present study ACA was preferred over AHP, because of its possibility for evaluating drivers simultaneously.

make trade-off decisions by estimating the impact on their quality management systems that they associate with particular drivers from their business environment, which are described in the next chapter. These drivers can be subdivided into levels: for example, the driver *media attention* may include the attribute levels *high* and *low* chance of negative *media attention*. In Appendix 3. the levels for each driver are described. Based on the attribute levels of the drivers, various business environments can be composed and the impact of each driver determined. In the context of the impact of drivers from the business environment on quality management, the basic assumptions are:
1. A business environment can be described by a set of drivers.
2. Expert judgements of the impact of the business environment on quality management are based on the combined choices of the levels of each driver.

In this study, the Adaptive Conjoint Analysis (ACA) is used which combines the design of conjoint tasks, data collection etc. in one piece of software. This approach compared to the paper based traditional full profile conjoint analysis, has a number of important differences, see Table 5.1 (Huber *et al.*, 1991). *Firstly*, the main difference (and advantage) of ACA above traditional full profile conjoint analysis is its computerised format which is customised (explaining the term 'adaptive' of ACA) to each respondent. This customised format means in practice that each respondent is asked in detail only about the drivers of greatest relevance to him (Van der Fels-Klerx *et al.*, 2000). Drivers of no interest for the respondent are neglected. As a result the ACA approach minimises the number of questions and time required to complete the interview, preventing fatigue bias. *Secondly*, in a full profile conjoint analysis the respondents are asked to determine their preference for alternatives on the basis of all attributes, specified at different levels. Respondents tend to use simplification strategies, because they are overloaded with many attributes each time (Orme, 2003). Therefore, the maximum number of attributes

Table 5.1. Differences between the Adaptive Conjoint Analysis and the Full Profile (adapted from Huber et al., 1991).

Adaptive conjoint analysis	Full profile
Combines design, data collection, analysis, and market simulation in one computer program	Design, collection, analysis and market simulation are n separated
Paired comparisons of profiles adapted to respondents prior evaluations	Usually the same set of full profiles for all respondents
Combines self explicated data with paired comparison intensity ratings	Fully decompositional approach
Can accommodate a large number of attributes	Restricted to about six attributes, unless bridging designs are used
Objects never fully specified (two or five attributes)	Objects specified on all attributes

in a traditional conjoint analysis should not exceed six (Orme, 2003). In the ACA method each of the alternatives is presented in partial profile, meaning that only a subset of the attributes is displayed for any given question (Huber *et al.*, 1991). As a result, ACA is able to measure more attributes than is advisable with the traditional full-profile method. *Thirdly*, in ACA the answers of each respondent can be checked for consistency and, if inconsistent, excluded from further analysis. *Fourthly*, after completion of the conjoint analysis, the estimated preferences of the respondent are directly available for discussion or analysis (Van Schaik *et al.*, 1998). *Fifthly*, Huber *et al.* (1991) have proved that ACA outperforms the full profile method by providing more consistent answer patterns.

5.1.2 Design

Within each conjoint analysis, the variables are ranked regarding a certain objective. In this study, variables were ranked according to their impact on quality management in agri-food supply chains. During the introduction of the ACA, the term quality management was explained (comparable with the elements of integrated quality management, explained in Section 3.2). In the present study, the ACA consisted of four steps comparable to the approach of Valeeva (2005), see Figure 5.2.

In the *first* step, the experts rated the impact of the different levels of drivers from the business environment on quality management systems using a seven-point Likert scale (1 = lower complexity; 4 no change; 7, higher complexity). The levels of the drivers are described in Appendix 3.

In the *second* step, the experts were asked to quantify the difference for the complexity[41] of quality management between the level with the lowest impact and the level with the highest impact on complexity of quality management for each driver on a seven-point Likert scale (ranging from 1 = very small difference, to 7, extremely big difference). These first two steps can be considered as the self explicated tasks in ACA. During these two steps, the experts were also asked to motivate their choices.

In the *third* step, experts were asked to compare several pairs of profiles and to judge which pair of profiles increases the complexity of quality anagement the most using a nine-point Likert scale (1 = strong preference for profile A, 9 = strong preference for profile B). A nine point scale was preferred, because, it allows the respondent to make more sophisticated differences. ACA can come up with scenarios which are very similar with regard to their impact on the complexity of quality management due to its adaptive nature (Sawtooth Software Inc). Holling *et al.* (1998) and also Lines and Denstadli (2004) found that profiles with two attributes make more accurate estimations than profiles with three or four attributes (preventing information overload on respondents). Therefore, in this study, the paired questions did not contain more than two attributes.

---

[41] Further in this study, the term impact is used instead of complexity.

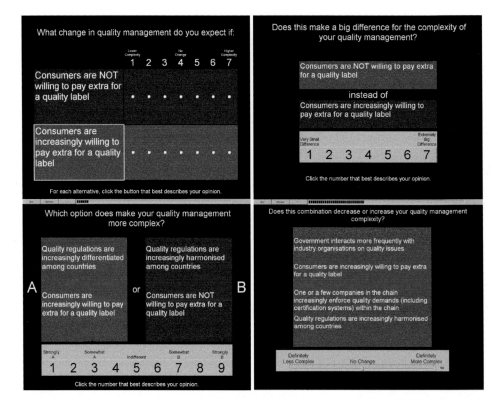

Figure 5.2. Steps of the adaptive conjoint analysis.

In the *fourth* step, customised profiles, or 'calibrating concepts' consisting of four drivers specified at certain levels were presented to the experts. These profiles were chosen to cover the whole range of drivers and their impacts on quality management. The profiles were presented one-at-a-time and the experts were asked to give their judgement as to what extent they think a profile would decrease or increase the complexity of quality management compared to the current situation on 0-100 points scale (0 = definitely less complex; 50 no change; 100 definitely more complex). For each respondent, the fit (R^2) between the likelihood score estimated by ACA and the actual scores of the calibration concepts was determined. This fit index was used as a measure for the consistency of the answer patterns of the respondents.

Before conducting the conjoint analysis, eleven experts pre-tested it. These experts came from the Product Board of Horticulture, the Dutch Food and Consumer Product Safety Authority, the EU-concerted Action Global Food Network and the Management Studies Group of Wageningen University. Three of the consulted experts were specialists in designing questionnaires and conjoint analyses, whereas the others were experts on the topics covered by this study.

5.1.3 Data collection and study population

Data were collected during the period January to April 2005. For an appropriate selection of experts, criteria were used such as the number of years of employment or experience in quality management, proven knowledge of quality management related topics, working in firms performing well on quality management (for business experts) or well-known research institutes (for research experts). Experts were asked to participate in the interview during a personal phone call explaining the objectives of the interview, which resulted in 47 appointments[42]. Furthermore, anonymity and confidentiality of the responses were assured. While answering the questions in ACA, each expert was asked to motivate his/her choices why he or she thought how specific levels of drivers would have an impact on quality management.

5.1.4 Data analysis

Firstly, the impacts of drivers from the business environment according to each individual expert were determined. Details of the regression layout and updating procedures in the ACA program are provided by Johnson (1987, 1991) and Green *et al.* (1991) and are not further described in this study. After this analysis, the relative impacts of drivers on quality management and the consistencies of the answers of each expert were available. Respondents showing a too low consistency would be removed from further analyses (no respondents were removed during the analysis). *Secondly,* tests were conducted to find out significant differences between the means of drivers between chains and between professional backgrounds, using the non-parametric Kruskal-Wallis test. The Kruskal-Wallis test is a one-way analysis of variance by ranks. It measures how much the group ranks differ from the average ranks of all groups. The only assumptions made by Kruskal-Wallis test are that the test variable is at least ordinal and that its distribution is similar in all groups (Siegel and Castellan, 1988). Therefore, this test is useful for comparing the means in small groups.

5.2 The questionnaire survey

The second phase in the present study was a large scale questionnaire survey. The questionnaire aimed to test the relationships in the research model aiming at answering questions such as: *How, and to what extent do several factors influence the integration of quality management? How does integration of quality management influence the dimensions of compliance behaviour?, and How do integration and self regulation of quality management influence performance of firms in agri-food supply chains?* This section describes the methodology employed to set up the questionnaire survey and to collect data to test the hypotheses developed in Chapter 4. An extensive review of the literature on the topics addressed in this study underlay the

---

[42] An ACA interview results in a set of individual customised set of attributes. Therefore, the minimum sample size is just one person Orme (1998). This suggests that if there was only one decision maker about the importance of the drivers one can learn a lot about that individual's preference from a conjoint analysis (Horst, 1996, Orme, 1998, Valeeva, 2005). An example of a conjoint analysis with only four respondents can be found in Hobbs (1996).

methodological operationalisation of the topics in the research model. This section further elaborates on the methods for assuring the validity and reliability of constructs, measuring the research topics. Finally, the section discusses the methods for testing the relationships in the research model.

5.2.1 Data collection and study population

Data were collected during the period from October to December 2005 by means of a paper based self administrative survey sent to the owners and the quality managers of the firms. This method was selected because the purpose was to examine patterns of associations, which requires quantifiable data and a large enough number of responses to allow for statistical testing. Little secondary data (data not gathered for the immediate study at hand) regarding the topics of the present study were available. The information that was available was often descriptive, fragmented or anecdotic in character.

Respondents make their decisions not only based on objective data, but also based on their subjective judgment (Clark *et al.*, 1994; Churchill, 1999; De Leeuw and Segers, 2002; Ketokivi and Schroeder, 2004). Although, previous studies have shown the questionnaire survey to be a viable research instrument for gathering such perceptual and subjective data (Claro, 2003; Lu, 2007), a lot of attention has to be paid to the survey design. After having established a proper research model, existing literature in which one or more elements of the research model were included, was searched for. The investigation of these surveys was helpful in obtaining validated questions that have been successfully applied in studies in the past. Also a lot of information obtained during the personal interviews accompanying the ACA was useful for the design of the questionnaire. After having developed a long gross list of questions, an iterative process was conducted in which questions were removed, or added, the formulation and sequence of questions were adjusted etc. After having obtained an acceptable number of questions that measured the elements of the research model well, a thorough pre-test was conducted.

The pre-test of a survey is one of the most critical success factors during its development (Snijkers, 2004). In the present study, expert review was used for pre-testing (Snijkers, 2002; Scheuren, 2004; Snijkers, 2004) using Tourangeau and Rasinski's (1988) modelled question-answering process, which consists of four steps; interpretation, information retrieval, judgement and rapportation. In each chain two potential respondents filled out the questionnaire[43]. *Interpretation* was concerned with the content validity of the survey, for example, are topics recognisable for the respondents, do the respondents know what the questions in the survey are aiming at, etc. *Information retrieval* is especially related to the efforts respondents face in answering the questions. For example, answering questions about topics, which have occurred a long time ago is difficult and will result in many missing values for such questions. Therefore, in this survey a maximum time period of three years was introduced for questions which

---

[43] Although this number might seem quite low, the questionnaires were comparable across the three chains in this study.

required information from the past. *Judgement* has to do with selecting the appropriate answer to the question, for example, how well does a certain statement apply to the situation of the respondent. *Rapportation* has to do with the ease of writing down the answers which was in most cases very simple, by indicating the appropriate score on the Likert scale. Pre-testing yielded all kind of information about content, formulation, scales, answer categories and layout which improved the survey a lot. After the pre-test the process of selecting potential respondents was started and after that the questionnaire survey was sent.

In order to make the survey more appealing recommendations of Dillman (1978; 2000) were applied. *Firstly*, an introductory letter was sent which explained the objectives of the study and the possible advantages of the outcomes of the study for the respondents. Name, address, telephone number and e-mail address of the researcher were included for possible enquiries. *Secondly*, to emphasise the interest of the present study for the chains involved, a recommendation committee was composed, consisting of:
- The Director of the Department of Industry and Trade of the Dutch Ministry of Agriculture Nature and Food Quality
- A main representative of the Department of Risk Assessment of the Dutch Food and Consumer Product Authority
- The chairman of the Product Board of Horticulture and the chairman of the Product Boards for Livestock, Meat and Eggs.
- Three well-known professors in the field of Supply Chain Management of Wageningen University and Research Centre.

Thirdly, the survey offered the respondent the possibility to add his or her name, address and e-mail address if respondents wanted to receive a summary of the most important findings from the survey[44]. *Fourthly*, on the survey there was also space left free for all kind of remarks regarding the study. *Fifthly*, a pre-paid envelope was included for the respondent to return the completed survey.

As was already stated in the introduction, this study was carried out in three chains (the poultry meat chain, the fruit and vegetable chain and the flower and potted plant chain) and in each chain two kinds of firms (primary producers and traders and/or processors) were selected, which means that in total six groups of firms were involved. For each of these groups, firms had to be selected. Guidelines on minimum sample sizes are not determined in literature and are dependent on the techniques used. However, estimation procedures have been found to provide valid results with sample sizes as small as 50, however, such a small sample size is not recommended. This study strives to gather at least 100 questionnaires per group, because according to the literature it seems that most statistical techniques work well with such a sample size. Expecting a response rate of 20% and including a 'safety margin', this implied

---

[44] The addresses were not provided to the researcher on a list, because privacy legislation prohibits that. The Industry Boards themselves printed the addresses on the envelopes. As a result it was not possible to retrace who had returned the survey.

that approximately 600 questionnaires had to be sent out for each group. The Product Board of Horticulture and the Product Boards of Livestock, Meat and Eggs were interested in the study and were willing to give support by providing addresses of potential respondents. In all three chains the largest firms were selected. These firms cover a much larger part of the total market, have often better future perspectives and have often more resources available for quality management compared to smaller firms.

In the poultry meat chain primary producers with more than 35.000 chickens were selected, which resulted in 599 firms. The total number of processors and traders (slaughterhouses and cutters) in the poultry meat chain was 313 and all these were included in the present study. For the fruit and vegetable chain, 600 vegetable growers with an area under glass of more than 10.000 square metres were selected. Furthermore, 600 traders of fruit and vegetables were selected. Also in the flower and potted plant chain 600 growers with more than 10.000 square metres under glass were selected and also 600 traders.

To minimise response bias, the knowledgeable respondent within each firm in terms of acquaintance with quality management was selected. For the traders and/or processors in the chains the survey was addressed to the employee who was responsible for quality management in the firm. For the primary producers, the questionnaire was sent to the owner of the firm. As a great number of primary producers are owner-managed, the owners are the informants who can provide relevant information for this study.

5.2.2 Measurement of the constructs

Many management studies seek statistical generalisability and are multi-industry investigations in which it is questionable to use operational definitions (for example number of employees, turnover, etc.). These problems have prompted researchers to rely on perceptual measures (Ketokivi and Schroeder, 2004). This study also uses perceptual and quasi perceptual measures for the constructs being investigated. Quasi perceptual means that something in the questions is compared to something else. An example of a quasi perceptual measure is: Compared to our main competitors, our revenues grow much faster or much slower. Further, Ketokivi and Schroeder (2004) argue that perceptual measures are viable alternatives in large sample studies, as long as rigorous examinations of validity are carried out. The present study uses a seven point Likert scale, which was preferred above a five point scale, (because including more items in the scale will increase its reliability (Churchill, 1999)). The constructs and the way they are operationalised are discussed below. Appendix 4 presents the items used in the questionnaire. All the constructs including their operationalisation, source and measures are summarised in Table 5.2.

External pressures

External pressures are the economic, technical, legal and ethical forces, which affect all firms, but on which the individual firms have less control (Omta, 1995). The drivers incorporated in the construct 'external pressure' were derived from the results of the ACA (see Chapter 6). The *first* driver *media attention* refers to the public exposure of business via television, radio, newspapers, magazines, films and books (Frombrun and Shanley, 1990; Greening and Gray, 1994). The *second* driver *legislative demands* refers to regulation by the state through the use of legal rules by sanctions (Black, 2002). The *third* driver *changing consumer demands* are the consumers' requirements to the characteristics of a product, process of service which satisfy their needs (Slack *et al.*, 1998). The *fourth* driver *societal demands for corporate social responsibility* refers to the inclusion of social and environmental concerns into corporate decision making and business operations as well as their interaction with stakeholders (Van den Brink and Van der Woerd, 2004). These drivers arc described in more detail in Section 6.2.

For computing the perceived pressure from the business environment, the two dimensions for measuring the impact of external drivers, *dependency* and *threat,* as described in Section 3.4 were used for each driver. Two questions measured the dependency on a driver by asking to what extent the driver was important for quality activities in the firm and to what extent the driver was important for competition in the market. For each driver the scores of the two dependency questions were averaged in order to calculate the overall dependency. A third question, measuring the level of threat for each driver was concerned with the negative impact of a driver on the revenues of a firm. The questions measuring both the dependency and the threat of drivers from the business environment were adapted from Klassen and Angell (1998).

The four environmental pressures, *media attention, legislative demands, changing consumer demands* and *societal demands for corporate social responsibility,* (see also Chapter 6) were combined into a perceived environmental pressure index representing the external pressure from the business environment as adapted from Fornell *et al.,* (1996).

$$\textit{External pressure index} = \frac{\sum_{i=1}^{4} w_i \bar{x}_i - \sum_{i=1}^{4} w_i}{6 \sum_{i=1}^{4} w_i} \times 100 \qquad \text{(Fornell \textit{et al.}, 1996)}$$

In this formula $\bar{x}_i$ is the value of the perceived dependency of driver i (unweighted average of the scores on the dependency questions) and w_i is the value of the impact of the driver on the revenue of a firm (the score on the threat question). The index for external pressure can range from 0 to 100, in which a high score means that a strong overall external pressure is perceived by the firm. This formula has been used in various contexts in which multiple variables determine the importance of an overall concept, for example, to measure consumer satisfaction in the United States (Fornell *et al.*, 1996), for predicting technology commercialisation success

Table 5.2. Operationalisation of construct, source and measures (see for exact wording of the measures Appendix 4).

Construct	Operational definition	Source	Measures
Factors influencing integration of quality management			
External pressure	The degree to which media attention, legislative demands, changing consumer demands and societal demands for corporate social responsibility have influenced quality management during the last three years	Omta (1995), Fornell et al., (1996), Sohn and Moon (2003), Klassen and Angell (1998), Kemp (2004)	Media1-3, Legis1-3, ConsDem1-3, CSR1-3
Transaction specific investments	Large investments made specifically for compliance with quality requirements of buyers with a large loss of value in case of redeployment/termination of the relationship (sunk costs)	Williamson (1975), Williamson (1985), Heide and John (1990), Buvik and Halskau (2001)	TsiSI-4/TsiCI-4
Information exchange by ICT	The use of information and communication technology (ICT) that connect separate organisations in both directions in the chain	Kumar and Van Dissel (1995)	ICTI
Quality strategy	The policy of a firm to establish, practice and lead a long-term vision for quality management in the organisation	Johnson and Scholes (1999), Rungtusanatham (2005)	ST1-3
Integration of quality management (second order construct)			
Monitoring	The gathering, analysis and evaluation of data in the chain for the compliance with quality requirements	Ahire et al. (1996), Forker et al., (1996), Forza and Fillipini, (1998), Buvik and Halskau, (2001),	MonSI-3/MonCI-3
Alignment	The strategic process of coordination of quality management systems of firms in the supply chain by means of long-term relationship development	Samson and Terziovski (1999), Forza and Fillipini (1998), Humphreys et al., (2004), Krause (1999) Flynn and Saladin (2001)	AlignSI-4/ AlignCI-4
Improvement	The management and use of information to maintain buyer focus and to improve quality performance of processes in the chain	Forza and Fillipini (1998), Samson and Terziovski (1999), Humphreys et al. (2004)	ImproSI-4/ ImproCI-4

Integration and self regulation of quality management

Self regulation			
Commitment	The degree to which firms know, accept and feel responsible to comply with quality requirements	(Morgan and Hunt, 1994), (Ministerie van Justitie, 2006), Ministerie van Justitie (2004)	ComSI-4/ ComCI-4
Enforcement	The ability of one channel member to influence the decisions and actions of another channel member by means of punitive actions	Hogarth-Scott and Daripan (2003), Ministerie van Justitie (2006), Ministerie van Justitie (2004)	EnfSI-3/EnfCI-3
Performance			
Buyer satisfaction	The degree to which a firm's buyer continually perceives that their (changing) quality needs are being met by the supplier's products and services	Rungtusanatham et al. (2005)	SatSI-3/SatCI-3
Revenue growth	The extent to which the revenues of the firm grows slower or faster compared to their main competitors	Flynn and Saladin (2001),	Grow1

(Sohn and Moon, 2003) and for calculating a perception of competition intensity of different industries (Kemp *et al.*, 2004)[45]. For more information about the mathematical details of the construction of this formula and a wide range of applications, the article of Fornell *et al.* (1996) is recommended.

Quality strategy

Quality strategy is defined as the policy of a firm to establish, practice and lead a long-term vision for quality management in the organisation (Johnson and Scholes, 1999; Rungtusanatham *et al.*, 2005). One of the major functions of the strategy of firms is to influence the setting of organisational values and develop suitable management styles to improve the integration of quality management in a firm's supply chain (Chen and Paulraj, 2004). For measuring the quality strategy questions from studies of Humphreys *et al.*, (2004), Chen and Paulraj (2004), Flynn and Saladin (2001) and Ahire *et al* (1996) were used. Questions about quality strategy referred to the support of firm's management for quality initiatives from inside and outside the focal firm, the priority of quality management in evaluating firm performance, the necessity of good quality management for the daily operations within the firm and the importance of good quality compared to the importance of price.

Transaction specific investments (TSIs)

TSIs are made specifically for the transaction with the selected counterpart with a large loss of value in case of redeployment or termination of the relationship (Williamson, 1985; Heide and John, 1990; Buvik and Halskau, 2001; Claro, 2003). As has been argued in Chapter 4., most TSIs have primarily centred on the human and physical dimensions of TSIs (Claro, 2003; Grover and Malhotra, 2003). Physical TSIs refer to investments such as equipment (or machinery), facilities, etc. specifically for transactions with the counterpart. The human TSIs refer to investments in human resources such as training of personnel in terms of dealing with specific requirements of a buyer. In the present study, no distinction was made between physical and human TSIs. The questions for TSIs were derived from studies of Buvik and Gronhaug (2000), Anderson and Buvik (2001), Buvik and Halskau (2001) and Krause *et al.* (1998) which have proved to be reliable scales. TSIs were operationalised by asking whether firms have largely invested in order to comply with quality requirements of buyers. Examples of such TSIs are investments in production means, information and communication structure, working routines (e.g. training) and adjustment of quality management systems. The adaptation of the structure of quality management systems between buyers and suppliers as an important TSIs came up during the interviews of the ACA.

---

[45] A driver could have a positive or a negative impact on quality management. However, by using this formula in which the threat of a driver to the firm's revenue is used as a weighting factor, a driver is converted into a pressure.

Information exchange by ICT

Information exchange by ICT is defined as the use of information and communication technology (ICT) to connect separate organisations in both directions (upstream and downstream) in the chain (Kumar and Van Dissel, 1995). More than ever before today's information technology is permeating the supply chain at every point, transforming the way exchange related activities are performed. In terms of quality management systems the use of ICT may be considered as an effective means for increasing the capacity of the exchange of information related to quality management. The use of ICT was operationalised with one question in which a firm was asked to what extent ICT was used for information exchange with other partners in the chain. The question was adapted from Chen and Paulraj (2004).

Integration of quality management

The second order construct of integration of quality management is composed by its three sub-constructs, representing the three dimensions *monitoring, alignment* and *improvement* as defined in Chapter 3 and refers to the term Supply Chain Quality Management as defined by Robinson and Malhotra (2005). For these dimensions the questions were formulated for the supplier and buyer model.

Monitoring is defined as to gather, to analyse and to evaluate data for compliance with quality requirements (Ahire *et al.*, 1996; Forker, 1997; Forza and Filippini, 1998). Integration of quality management means that specific systems have to be introduced within supply chain relationships that can handle inter-firm coordination, implementation of monitoring and control of actions to assure quality. Within these inter-firm relationships specific tasks and responsibilities between firms have to be planned in detail (Buvik and Halskau, 2001). Questions for measuring monitoring deal with issues such as the transfer of outcomes of specific quality tests and - inspections and active participation in the monitoring systems of buyers. The items for measuring monitoring were adapted from studies of Buvik and Halskau (2001), Buvik and Gronhaug (2000) and Flynn and Saladin (2001).

Alignment is defined as the adjustment of each others quality management systems of firms in the supply chain by means of long-term relationship development (Benton and Maloni, 2005). Firms which expect the relationship to last for a long time may be more willing to engage in the development of win-win relationships and to make quality management more supply chain oriented (Krause, 1999; Robinson and Malhotra, 2005). The items for measuring alignment were based on studies of Samson and Terziovski (1999), Humphreys *et al.* (2004) and Krause (1999). Questions in the alignment construct ask to what extent tight collaboration, special appointments and the communication of quality requirements are made between a firm and its suppliers and buyers.

Improvement is defined as the management and use of information to maintain customer focus and to improve quality performance of processes in the chain (Samson and Terziovski, 1999). To effectively measure this construct, the study used questions which were successfully used in the study of Samson and Terziovski (1999) and Humphreys *et al.* (2004). Items measuring improvement deal with the extent to which feedback about quality performance is transferred back in the chain, the transfer of quality requirements of customers upstream in the chain and the measurement of buyer satisfaction related to the quality performance of other chain partners.

Self regulation

Self regulation is based on the 'Table of Eleven' of the Dutch Ministry of Justice and contains two dimensions, commitment and enforcement. *Commitment* can be defined as an exchange partner's (for example, a supplier or buyer) belief that the relationship is worth working on to ensure that it endures indefinitely (Morgan and Hunt, 1994). As was already argued in Chapter 3 commitment in the 'Table of Eleven' focuses on affective commitment and not on continuance and normative commitment. In this study, it is assumed that due to commitment suppliers become loyal to the specific quality requirements of buyers. Not all the dimensions of commitment from the 'Table of Eleven' are included in the survey, because the 'Table of Eleven' is in fact an expert tool in which an expert judges the behaviour of a certain group. Questions that measure commitment to quality requirements were for example, the extent to which people were familiar with quality requirements of their buyers, found them reasonable and felt responsible for them. Commitment was measured for both the supplier and buyer model.

Enforcement is a form of coercive power and can be defined as the ability of one channel member to influence the decisions and actions of another channel member by means of punitive actions (Hogarth-Scott and Daripan, 2003). In the 'Table of Eleven', six dimensions for enforcement exist, which deal with the frequency of control and sanctions at a very detailed level. In order to prevent confusion among the respondents, enforcement was measured as the frequency of control and severity of sanctions in case of non compliance. Enforcement was measured for both the supplier and buyer model.

Performance

The literature on performance indicators shows that the *sales growth rate* and *revenue growth* are the financial performance indicators most commonly used (Mohrman *et al.*, 1995; Beamon, 1999).

Revenue growth – Regarding the different sizes of the firms and different chains included in this study, *revenue growth compared to competitors* turned out to be the best indicator for measuring the financial performance. Revenue growth uses a quasi perceptual scale by asking to what extent the revenues of the firm grow slower or faster compared to their main competitors.

The question for revenue growth compared to the main competitors was based on studies of Flynn and Saladin (2001) and Conca *et al.* (2004) and was only measured for the focal firm. Sales growth rates was only used for description purpose.

Buyer satisfaction – This performance indicator is supported by the notion that a firm's performance is determined in part by how well the relationship fulfil expectations (Claro, 2003). In this study, buyer satisfaction measured satisfaction of a buyer with regard to the quality of the products and processes of a supplier and the ability of the supplier to respond rapidly to changing quality requirements of the buyer. Most of these questions were derived from the study of Flynn and Saladin (2001). Again for this construct the questions were formulated for the supplier and buyer model.

Control variables

In this section variables are described that were expected to have an impact on the elements of the research model. Previous research suggests that buyer-supplier relationships in agri-food supply chains might be affected by the presence of a chain leader, the number of quality management systems, size of the firm, the presence of a quality manager, age of the respondents and the number of suppliers and buyers.

A *chain leader* can be defined as a dominant firm in the supply chain which possesses superior negotiation power. It imposes its strategy and objectives on the other firms in the supply chain (Lejeune and Yakova, 2005). For example, a chain leader may specify the quality requirements of products and processes and the control mechanisms to be enforced to its suppliers (Humphrey and Schmitz, 2002). The presence of a chain leader was measured by whether or not respondents recognise a firm in their chain that was able to enforce its quality requirements on their firm.

Usually, compliances to standard quality systems are considered as direct measures available for communicating quality performance requirements to suppliers and buyers. They are often accepted and audited industry standards. These systems guarantee basic levels of quality assurance especially for commodity goods (Simpson *et al.*, 2005). More quality conscious buyers may desire higher levels of quality for performance of their suppliers that go beyond traditional quality management systems by requiring compliance with *extra quality management systems*. In the survey, respondents could indicate with which quality management systems they were currently complying.

Size of the firm is an important control variable in this study. Smaller firms have flatter organisational structures and more informal communication channels. As a result quality management systems may be more effectively implemented in small firms. On the other hand, larger firms have more market power, capital resources and professional and managerial experience. Authors such as Taylor and Wright (2003) and Powell (1995) found that firms that have discontinued quality management were predominantly small in size. However, Ahire and Dreyfus (2000) and also Ahire and Golhar (1996) did not find any impact of size on quality management. The size of the firms was measured by the number of employees of a firm and the yearly turnover.

It is expected that the presence of a *quality manager* within a firm will have an impact on the level of integration of quality management. It is supported by literature that an experienced, technically qualified person is an important factor influencing the implementation of quality management systems (Taylor, 2001; Esbjerg and Bruun, 2003).

Age is an important control variable for primary producers. In some cases older owners might have more experiences and knowledge. However, it could be expected that younger owners have more experience with modern communication means and are more computer literate than older farmers, which may be advantageous in today's agri-food supply chains with its strong dependency on information sharing. Only primary producers were asked to write down their age in the questionnaire survey. For traders and/or processors age is less relevant. These firms are larger and have more personnel in order to continue the firm. Furthermore, when primary producers foresee that their firm will finish its activities; this might have an impact on the integration of quality management. For instance the development of long-term relationships will become less important. Therefore, the *presence of a successor* was also measured for primary producers. These two variables were combined into one control variable measuring whether or not an owner was young and/or had a successor.

The number of *suppliers and customers* are important control variables for only traders and/or processors. Since collaboration becomes an important issue in supply chain management new competition insights suggest close long-term working relationships with only a few partners (Krause, 1999; Green *et al.*, 2006). Traders and/or processors were asked in the questionnaire to write down the number of suppliers and buyers. Also the question was added to which kind of buyer they were delivering their products.

5.2.3 Assessing the validity and reliability of constructs

This section discusses the assessment of the validity and the reliability of the *reflective* and *formative* constructs used in this study. In Appendix 1 the definition and comparison of formative and reflective constructs are discussed in detail. Formative constructs have characteristics that are different compared to the reflective constructs. Literature indicates that three issues are critical for formative constructs: content validity, nomological validity and item multicollinearity. The validity and reliability of reflective constructs were assessed by following the procedures described by Anderson and Gerbing (1988) and Steenkamp and Van Trijp (1991). For reflective constructs, content validity, convergent validity, discriminant validity and nomological validity are important validity procedures. Procedures conventionally used to assess the validity of reflective constructs are factor analysis (both explorative and confirmatory) and item-total correlation. Furthermore, it is needed to asses the reliability of the reflective constructs, which is assessed by means of Cronbach's α, composite reliability and variance extracted. In Table 5.3 the threshold levels of the evaluation criteria for the validity and reliability of the reflective constructs are summarised. The difference as well as the methods used to assess formative and reflective constructs are also described in detail in Appendix 1.

5.2.4 Quantitative methods for data analysis

In this section the methods for hypothesis testing are presented. In Chapter 3 and 4 the research model and the hypotheses have been presented. By estimating structural equation models, the present study tested the hypotheses. Before developing an overall model for all the groups of firms included in this study, the generalisability of the model across the groups was tested, using multi-group CFA and multi-group structural equation modelling (SEM). The effects of (group) specific control variables were also investigated, which increased the number of estimated parameters considerably. Testing the effect of control variables would add many non-significant paths to the structural model resulting in a decrease of fit. In addition, some control variables are binary variables, which make them less suitable for SEM which is very sensitive to deviations of multi-normality (Hair *et al.*, 1998). Finally, some control variables were not measured in each sub-group. Therefore, multiple regression equations to analyse the effects in each sub-group were used instead of SEM.

Table 5.3. Overview of the statistical evaluation criteria for reflective constructs.

Evaluation criteria	Threshold
Validation of construct	
Inter-item total correlation	≥ 0.50
Explorative factor analysis	
Explained variance	$\geq 60\%$
Factor loadings	≥ 0.60
Confirmatory factor analysis	
Standardised loadings (λ)	≥ 0.60
t-value of the standardised loadings	≥ 1.96
Reliability of the constructs	
Cronbach's α	≥ 0.60
Composite reliability	≥ 0.70
Composite validity (variance extracted)	≥ 0.50

Multiple regression

Multiple regression is likely the most frequently used technique for analysing interdependence relationships. It is used to explore the relationship between a number of independent variables and a single dependent variable. Multiple regression analysis can be used for quantifying the best relationship between a dependent and a number of independent variables. Multiple regression also offers the possibility to determine which of the independent variables has the strongest relationship with the dependent variable and what the direction (positive or negative) of the relationship is. An important thing to note is that especially in social sciences relationships between the dependent and independent variables are in most cases not perfect. Many unobserved variables may interact with the independent variable.

The regression model is a linear combination of independent variables that corresponds as closely as possible to the dependent variable (Lattin *et al.*, 2003). In a two dimensional example this means that regression analysis estimates the line of best fit by minimising the vertical distances between the points used to estimate the line. The line of best fit is called the regression line. The vertical distances between the points and the estimated lines are squared and used as a measurement of the total sum of error. In fitting the line, the ordinary least squares procedure minimises the sum of the squared errors. A general multiple regression equation has the following form:

$$Y = \beta_0 + \beta_1 X_1 + \beta_2 X_2 + \beta_3 X_3 \ldots\ldots\ldots\ldots \beta_k X_k + \varepsilon \qquad \text{(Hair *et al.*, 1998)}$$

In this equation Y is the dependent variable and the X's are the independent variables and ε is the error. β_0 is the intercept of the regression line. The coefficient β_k is the relative contribution of the independent variable k to the overall prediction of the dependent variable and represents the standardised partial regression coefficient. ε is the error term of the prediction (Churchill, 1999).

Regression coefficients are expressed in terms of the unit of the variables, thereby making comparisons between coefficients inappropriate. However, the standardised β coefficient enables an evaluation of the relative effect of each independent variable on the dependent variable. The standardisation process transforms the absolute regression coefficients into a new coefficient with a mean of 0 and a standard deviation of 1. By this transformation, the β_0 term (the intercept) turns into the value 0. Thus, β coefficients use standardised data and can be directly compared. The significance of the β_k is assessed by the t-values which have values of 1.645 for 10% significance level; 1,960 for 5% significance level and 2.326 for 1% significance level, in case of two tailed t-tests.

Multiple regression also provides outcome R^2, which is called the coefficient of determination, one of the most important measures. This coefficient represents the proportion of variation in the dependent variable that is accounted for by the co-variation in the predictor variables. The adjusted coefficient R^2 takes into account the number of independent variables and the sample size. This measure gives an insight into what extent certain independent variables significantly influence the dependent variable. Usually R^2 rises if more predictor variables are included in the model, but if these predictor variables have hardly any significant contribution to the explanation of the dependent variable, R^2 adjusted will remain almost the same. R^2 (adjusted) ranges from 0 to 1 and the higher the value, the better the explanatory power of the regression equation. The significance of R^2 (adjusted) is assessed by the magnitude of the F statistics.

Another important item in multiple regression is the Variation Inflation Factor (VIF) which can be used to assess multicollinearity within the data. If multicollinearity is present, the independent variables are highly correlated and are interchangeable. When a multiple regression is carried out with variables showing high levels of multicollinearity it will not be clear which variable accounts the most for the variation in the dependent variable (Field, 2003). If the largest VIF is greater than 10 there is a concern for multicollinearity. The VIF is directly related to another measure of multicollinearity, the tolerance value (TV), in the following way:

$$VIF = \frac{1}{TV} \hspace{4cm} \text{(Field, 2003)}$$

Assuming that multicollinearity is a problem if the VIF exceeds 10, the tolerance value should be larger than 0.10 (Hair *et al.*, 1998; Field, 2003).

Structural equation modelling (SEM)

SEM combines aspects of multiple regression (examining dependence relationships) and CFA (representing the construct part of the variables) to estimate a series of interrelated dependence relationships simultaneously. The most important characteristics of SEM which make it different from other techniques are the parallel estimation of multiple and interrelated dependence relationships, the ability to represent unobserved concepts in these relationships and account for measurement error in the estimation process. SEM is a powerful method for testing causal models, because it provides the total effects (i.e. direct and indirect effects) and the complete model's goodness of fit. In fact SEM carries out a series of separate, but interdependent, multi regression equations simultaneously, by specifying the structural model used in the statistical program Lisrel (Hair *et al.*, 1998).

The structural model is derived from the research model. The theoretical concepts are operationalised in a set of observed variables (e.g. scales and indicators) that are later computed into latent variables. Thus, by using this technique it is possible to test a structural model between variables that reproduce the influence of latent independent variables on latent dependent variables.

Box 5.2. Goodness of fit indices.

After estimating the measurement or structural model, given a converged and proper solution, an assessment is needed on how well the specified model accounts for the data. This is done with one or more fit indices. These indices determine the degree to which the model predicts the observed covariance matrix (Hair *et al.*, 1998). They are the *Chi-Square* (χ^2), the χ^2/df (degrees of freedom) statistic, *the goodness of fit index (GFI), the adjusted goodness of fit index (AGFI) and the Root Mean Square Error of Approximation (RMSEA)*. Incremental indices should also be used, such as the *normed fit index (NFI), the non-normed fit index (NNFI), the comparative fit index (CFI)* and *the Consistent Akaike Information Index (CAIC)*.

If the proposed model fits well with the observed data, the χ^2 will be non-significant and its value should fall between two or three times the number of df (Tabachnick and Fidell, 2001), although for larger samples the proportion might be 8. A large value of χ^2 relative to the df signifies that the observed and estimated matrices differ considerably. Statistical significance levels of χ^2 indicate the probability that these differences are caused solely by sampling variation. Unfortunately $\chi2$ is not very useful for this particular purpose, because of its sensitivity to sample size, model complexity, problems of trivial fit and improper solutions (Fornell, 1983).

The *RMSEA* is a measure of discrepancy between the reproduced and observed covariances per degree of freedom. Values of 0.08 and below indicate acceptable fit (Tate, 1998). The *GFI* and the *AGFI* both assess how much better the proposed measurement model fits the data as compared to no model at all. The value of both the GFI and the AGFI can range from 0 to 1 where a high value means a better fit. Hair *et al.* (1998) recommend a value of 0.90 or greater for the GFI and 0.80 for the AGFI.

The *CFI* compares the existing model fit with a null model which assumes the latent variables in the model are uncorrelated (the 'independence model'). That is, it compares the covariance matrix predicted by the model to the observed covariance matrix, and compares the null model (covariance matrix of 0's) with the observed covariance matrix, to gauge the percent of lack of fit which is accounted for by going from the null model to the researcher's SEM model. CFI varies from 0 to 1. CFI close to 1 indicates a very good fit. By convention, CFI should be equal to or greater than 0.90 to accept the model, indicating that 90% of the covariation in the data can be reproduced by the given model (Tabachnick and Fidell, 2001).

The *NFI*, an incremental index for the χ^2 statistic, was developed as an alternative to CFI, but one which did not require making chi-square assumptions and does not penalise for sample size. It varies from 0 to 1, with 1 = perfect fit. NFI reflects the proportion by which the researcher's model improves fit compared to the null model (random variables). For instance a NFI of 0.50 means the researcher's model improves fit by 50% compared to the null model. By convention, NFI values below 0.90 indicate a need to respecify the model (Tabachnick and Fidell, 2001).

The *NNFI*, an incremental index for the χ^2/df, also called Tucker-Lewis index (*TLI*), is similar to NFI, but penalises for model complexity as reflected in the degrees of freedom of the independence and research models. NNFI is not guaranteed to vary from 0 to 1. NNFI close to 1 indicates a good fit. By convention, NNFI values below 0.90 indicate a need to respecify the model (Tabachnick and Fidell, 2001).

The *CAIC* is an incremental goodness-of-fit measure which adjusts model chi-square to penalise for model complexity and sample size. Thus, CAIC reflects the discrepancy between model-implied and observed covariance matrices. CAIC close to zero reflects good fit and between two CAIC measures, the lower one reflects the model with the better fit (Tabachnick and Fidell, 2001). For evaluating the fit of a model, researchers should never solely rely on one fit index, but evaluate them simultaneously.

Multi-group confirmatory factor analysis and structural equation modelling

As already described in Section 1.4 in this study, three chains were included, while in each chain two groups of firms were included, primary producers and processors and or traders, resulting in six different groups of firms. For studies in which data form different independent samples was analysed, a serious limitation is often the assumption that the set of items and the number of underlying constructs has to be the same across all samples (Baumgartner and Steenkamp, 1998). If between groups of firms meaningful comparisons should be made, the measurement model has to be the same (or invariant) to a certain extent for all groups of firms. Otherwise, finding differences between structural models of different groups are open to analogous different interpretations (Steenkamp and Baumgartner, 1998). In case of non-invariant models significant differences within the structural model can be a 'true' significant difference, but can also be the result of different measurement models. Therefore, a multi-group analysis in Lisrel was performed, in which data from the six independent samples is analysed simultaneously. This is an appealing methodology for testing measurement equivalence and for investigating invariance hypotheses of substantive interest such as whether construct means are equal across groups or whether the magnitude of a structural relationship is the same in different groups, etcetera (Baumgartner and Steenkamp, 1998).

Steenkamp and Baumgartner (1998) have argued that if the purpose of a study is relating constructs to other constructs in a structural model, metric invariance has to be satisfied, because the scale interval of the constructs has to be comparable across the different samples. Metric invariance means that the factor loadings on the constructs do not differ significantly across the groups (thus, for example, λ_1 of construct A in group 1 is not significantly different from λ_1 of construct A in group 2, see Figure 5.3). In addition to metric invariance, factor[46] variance invariance is required, because comparisons of standardised measures of association (such as correlation coefficients and standardised regression coefficients) will be made within this study. Factor variance invariances means that the variances of the constructs do not differ significantly across the groups (thus, for example, ξ_A of construct A in group 1 is not significant different from ξ_A of construct A in group 2).

---

[46] For factor, also the word construct could be used.

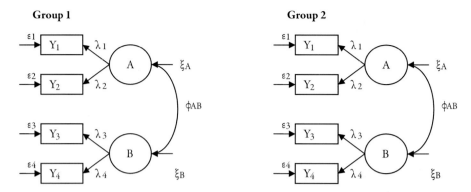

Figure 5.3. An example of multi-group confirmatory factor analysis.

To further improve the results the factor covariance invariance was also tested. More stringent invariant models are always preferred, because they further strengthen the conclusions. Covariance invariance implies that co-variances between the constructs in the groups of firms do not significantly differ and that the paths in the structural model for each group are not significantly different (thus, for example, ϕ_{AB} between construct A and B in group 1 is not significant different from ϕ_{AB} of construct A and B in group 2, see Figure 5.3).

According to Steenkamp and Baumgartner (1998), lack of error variance invariance is not required, as long as differences in measurement errors are explicitly taken into account which is the case of latent variable modelling. However, the scale reliabilities should be about the same when measures of association between observed variables are compared between the firms. In order to test for metric and factor (co) variance invariance for the factors, the latent constructs have to be assigned in the scale in which they are measured. For comparing different groups this is usually done by setting the factor loading of one item to one per construct. These items are referred to as marker (or reference) items. Within each group, the same items should be used as marker items (Steenkamp and Baumgartner, 1998).

In addition, Baumgartner and Steenkamp (1998) have described a procedure for testing several kinds of invariances. Models should be represented in nested models so that systematic model comparisons can be made. A well known assessment criteria for investigating invariance is the model comparisons with the Chi-Square Difference Test. However the Sequential Chi-Square Difference Test suffers from the same problems as the Chi-Square test for evaluating the model fit, such as sensitivity to sample size and model complexity. With large sample sizes, significant values can be obtained even though there are only trivial discrepancies between a model and the data and the same holds for more complex models (Laros and Steenkamp, 2004). Therefore, Steenkamp and Baumgartner (1998) and also Anderson and Gerbing (1988) recommended using and placing more emphasis on the following fit indices: RMSEA, CAIC, CFI, NNFI. Higher values of the NNFI and CFI and lower values of the RMSEA and

CAIC indicate better models. Steenkamp and Baumgartner (1998) stress that for comparison of models especially RMSEA, NNFI and CAIC seem to be particularly useful because they take into account both goodness of fit and model parsimony[47], by imposing a penalty on fitting additional parameters. These fit indices and especially the CAIC were found to be the most effective indices in distinguishing between correctly and incorrectly specified models (Steenkamp and Baumgartner, 1998).

After having validated the required invariance of the measurement model, the next step in testing the invariances is testing the invariance of the paths (comparisons of the gammas and betas in the structural models) across the six sub-groups. Paths can be regarded as the proposed hypotheses in the research model and are described by a standardised coefficient β, (or γ for the effect of an exogenous variable on an endogenous variable) with a specific t-value, see Figure 5.4.

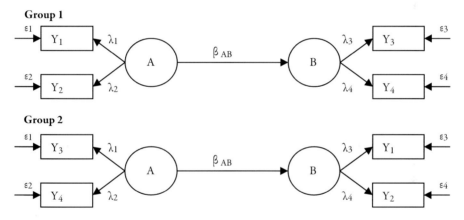

Figure 5.4. Testing for path invariance.

---

[47] Philosophers of science have long argued that the objective of science is not only to explain, to predict and to understand the world in which they live, but to do so in as efficient manner as possible (Morgan and Hunt, 1994). Parsimony or 'simplicity' has been so important in science. When certain hypotheses are equally satisfactory in other respects the researcher should choose the simpler. The simpler hypothesis is usually the more elegant, more convenient to work with, more easily understood, remembered and communicated. The emphasis on parsimony in the structural equation modelling literature is fully in accord with science (Bentler and Mooijaart, 1989). Parsimony goes back on Ockham's razor which stated that for the explanation of a certain phenomenon in as few assumptions as possible should be made, by eliminating or 'shaving off' those assumptions that make no difference in the observable prediction of the explanatory hypothesis or theory. As a result the less complicated alternative should be chosen, the lex parsinoniae (law of succinctness):*entia non sunt multiplicanda praeter necessitatem, (entities should not be multiplied beyond necessity).*

This β or γ is comparable with the β in multiple regression. Although the name 'standardised coefficient' suggests the opposite, standardised coefficients can be larger than 1 (Jöreskog, 1999). Invariance of the paths was tested in the following way, see also Figure 5.4:

1. A constrained model was tested with equality constraints for all the path coefficients across all the groups and the same kind of reasoning as for the measurement model was followed for evaluating the obtained solutions. Thus, β_{AB} in group 1 is the same as β_{AB} in group 2.
2. The equality constraints for each path were set free one at a time and the fit indices were compared with the fit indices of the fully constraint model (Ahire and Dreyfus, 2000). Thus, β_{AB} in group 1 is allowed to be different from β_{AB} in group 2.
3. A final test was performed in which the totally constrained model was compared with the model in which all equality constraints of the paths were set free all together.

For evaluating the outcomes the same indices as for testing the invariance of the measurement model are used. If the measurement model and the structural model can be constrained over all the groups no detailed multi-group analysis is needed (Laros and Steenkamp, 2004). In that case, all data can be pooled together and treated as one group.

5.3 In-depth interviews

This section describes how the research model is applied in a number of in-depth interviews, the *third* phase in the present study. Conducting in-depth interviews at the end of the research project is a desirable strategy, because by focusing on the 'how' and 'why' questions, the in-depth interviews help to better explain the relationships found and to add practical insights to the survey results (Melnyk and Handfield, 1998; Meredith, 1998). The in-depth interviews were especially aimed at answering the fourth research question in Chapter 1:

> *What is the best way to create self regulated quality management systems in agri-food supply chains?*

5.3.1 Design

Because the number of in-depth interviews in many studies is small and the data is rather subjective, the weakest point often mentioned is that generalisability or external validation is under pressure (Verschuren and Doorewaard, 1999). In order to increase the validity and reliability of in-depth interviews the use of a theoretical basis and asking respondents to review the interview results are strongly recommended (Yin, 1994; Verschuren and Doorewaard, 1999). The present study has used the research model developed in Chapter 3 as a guideline during the in-depth interviews. For each of the concepts and the relationships between them, evidences and explanations from the in-depth interviews were gathered. Furthermore, the questions in the in-depth interviews focused on topics such as: the distinctive characteristics of 'best practice' quality management systems compared with 'normal' quality management systems, success factors and advantages and disadvantages of 'best practice' quality management systems.

The in-depth interviews also paid a lot of attention to self regulation. The questions about self regulation discussed topics such as: The role of the government and other organisations such as the industry organisations, certifying organisations and interesting organisations in designing self regulated quality management systems, critical success factors and advantages and disadvantages of self regulation for quality assurance in agri-food supply chains.

The questions were open-ended, because then the respondent would not be hindered by any framework or bias of the researcher. The questions being used in the in-depth interviews are described in Appendix 4.

5.3.2 Data collection and study population

The semi-structured interviews were held from June to October 2006. For selecting respondents, the same criteria as for ACA described in Section 5.1 were used. Because the objective of this part of the study is to get more insight in the 'best practices' in quality management, an important criterion was that respondents have to be employed at firms that had 'best practice' quality management systems operational. In order to find firms with such systems, three approaches were combined:
1. At the end of the questionnaire, respondents were able to fill out their name in order to obtain feedback on the results or to participate in future research. Based on the scores on the constructs measuring integration of quality management, a number of firms were selected.
2. After selecting some firms, the Internet sites of the firms were explored in order to retrieve information about the way quality management was carried out.
3. During the conjoint analysis, a number of firms were already identified with 'best practice' quality management systems.

The aim was to interview at least one primary producer and one trader and/or processor in each chain. In order to increase the generalisability of the in-depth interviews for each chain experts from interest organisations (e.g. Product Boards or trade associations) were also interviewed. These experts were directly involved in the development, control and enforcement of quality management systems. It was expected that these respondents would have a 'helicopter' view over their sector, enabling them to make general statements and revealing the difference between firms with and without 'best practices'. In further improving the generalisability representatives from certification organisations were included because their experience with 'best practice' quality management enabled them to make appropriate comparisons between firms. Finally, one person from the Ministry of Agriculture, Nature and Food Quality and one person from the Food and Consumer Product Authority (VWA) were interviewed in order to get an insight in the opinion of governmental organisations on 'best practice' quality management and self regulation. Anonymity and confidentiality of the experts were assured.

A couple of days before the interviews the respondents received a questionnaire. Each interview was tape-recorded and took on average one and a half hours. The transcripts of the interviews were analysed and a case report was written for each firm. The respondents received a copy of the written reports and were asked to comment on it in order to enlarge the validity of the results. Data analysis was conducted by reading and comparing the case reports.

5.4 Concluding remarks

This chapter described the methods for the three phases in this study: conjoint analyses, questionnaire survey and in-depth 'best practice' interviews.

In the *first* phase a conjoint analysis aimed at ranking the most important drivers from the business environment. The motivation of the respondents for their choices helped to refine the thoughts about the relationships between the business environment and the integration of quality management. This information was extremely helpful for constructing the second and most important phase of the study.

In the *second* phase of the study, a questionnaire survey enabled the relationships to be quantitatively tested. The operationalisations of the items measuring the constructs were discussed as well as the validation and reliability procedures for reflective and formative constructs and the methods to be used to analyse the data.

The *third* and last phase of the study consisted of a number of in-depth interviews in which the main findings of the questionnaire survey were validated. The outcomes of these interviews are used for formulating practical recommendations for policy makers and managers to build self regulated 'best practice' quality management systems in agri-food supply chains.

This 'mixed methodology' approach is expected to increase the 'goodness' of the answers to the research questions (Teddlie and Tashakkori, 2003). The outcomes from the three phases are discussed in the Chapters 6, 7 and 8.

Chapter 6. Conjoint analysis results

This chapter presents the outcomes of the Adaptive Conjoint Analysis (ACA). The conjoint analysis was the first phase in this study and was aimed at the identification and ranking of drivers from the business environment which have an impact on quality management systems in agri-food supply chains. The chapter starts with a brief discussion of the characteristics of the experts interviewed in Section 6.1. It further continues with a description of the most important drivers from both the general and the task environment in Section 6.2. The ranking of the drivers according to chain and professional background is discussed in Section 6.3. The chapter ends with a number of concluding remarks in Section 6.4.

6.1 Study sample

The Adaptive Conjoint Analysis (ACA) was conducted in the period from January to April 2005. Willingness to participate in an interview was asked in a personal phone call in which the objectives of the interview were explained. 47 experts agreed to participate in ACA, 30 experts from business and 17 experts from research. They received a personal visit at their working addresses. Research experts were employed at organisations such as TNO[48], the Food and Consumer Product Authority, the Institute of Food Safety, the Ministry of Agriculture, Nature and Food Quality and Wageningen University and Research Centre. Experts from business were employed at firms, interest organisations, Product Boards, certification firms and trade associations. Table 6.1 shows the distribution of the professional background across the three chains.

Within the flower and potted plant chain, research institutes dealing with quality management were scarce compared to the other two chains, explaining the relatively low number of research experts involved in this chain. Experts were asked to motivate their choices on the impact

Table 6.1. Distribution of professional background among the chains.

Experts from:	Poultry meat	Fruit and Vegetables	Flowers and potted plants
Business	8	10	12
Research	7	7	3

---

[48] Nederlandse Organisatie voor Toegepast Natuurwetenschappelijk Onderzoek, in English: Netherlands Organisation for Applied Scientific Research.

of the drivers from the business environment on quality management during ACA. All the interviews were tape-recorded and were transcribed in a case description. To increase quality and reliability of the present study, the experts reviewed the transcripts and their individual ranking of the drivers and amended them if necessary. All experts accepted their outcomes and the transcripts, implying that the reliability of the study was effectively addressed.

6.2 Drivers

For the identification of drivers from the business environment which have an impact on quality management in agri-food supply chains an extensive literature study and a number of in-depth interviews with experts involved in the EU concerted action 'Global Food Network' have been carried out. Based on the literature study and the interviews twelve drivers were derived from the general and task environment (see Section 3.4) which are shown in Table 6.2 and are discussed below[49].

Table 6.2. Selected drivers from the general and task environment included in the conjoint analysis.

General environment	Task environment
Media attention	Increasing power dependency in the chain
Supra- and national legislative demands (two separate drivers)	Chain wide innovation of quality management systems
Changing consumer demands	Information exchange by ICT
Societal demands for corporate social responsibility	
Willingness to pay for a quality label	
Globalisation of import and export (two separate drivers)	
Different quality regulations/systems	

6.2.1 Drivers from the general environment

Media attention

Media attention can be defined as the public exposure of business via television, radio, newspapers, magazines, films and books (Behr and Iyengar, 1985; Frombrun and Shanley, 1990; Greening and Gray, 1994). Mass media and specialised publications propagate an evaluation of firms' activities by the public, which is especially true for firms operating in

---

[49] Information exchange in the chain by means of ICT was extensively discussed in the previous chapters.

controversial product-market domains (Frombrun and Shanley, 1990). Crises in agriculture such as the BSE crisis, dioxin crisis, classical swine fever and foot and mouth disease (see Box 2.1) have generated a considerable amount of negative media attention to food production in Europe (Lloyd *et al.*, 2001; Verbeke and Viane, 2002; Trienekens and Van der Vorst, 2003). As a result, consumers started to rethink their attitude with regard to agri-food products. Due to media attention the consumption level of agri-food products which are involved in food crises has dropped heavily on several occasions. For example, Verbeke and Ward (2001) found a negative press/advertising ratio of five for beef consumption which means that the total beef expenditure gain attributed to advertising is five times lower than the loss resulting from negative publicity. Therefore, firms in agri-food supply chains have a common interest in the integration of quality management: if consumers lose confidence, this affects all firms in the chain (Giraud-Héraud *et al.*, 2002; Mazé, 2002; Grievink *et al.*, 2003).

Legislative demands (national and supra-national demands)

Legislative demands have been identified by many authors as the most important driving factor for implementing quality management systems (Downey, 1996). However, the managerial impacts of legislation can vary dramatically as was shown in the case of environmental management systems by Klassen and Angell (1998). Due to the huge interest of society in safe and high quality food products, the food industry has become a heavily regulated industry. Firms that do not comply with legislative quality demands are subject to a range of penalties, including fines, product recalls and temporary or even permanent restrictions to their production (Henson and Hooker, 2001). Through the judicious use of incentives, governments are often able to structure a market, so that markets fulfil public purposes. In this study a distinction is made between supranational (demands from the EU and outside the EU) and national legislative demands. The explanation was that European legislative demands are transferred into national legislation, often on a more detailed level. Regarding national legislative demands, experts were asked to indicate the impact of decreasing or increasing governmental interaction on quality issues in their industry. Interaction between the government and the food industry is increasingly important for the compliance process (Donker *et al.*, 2000). In this way the government and industry create common goals ensuring that foods are safe and provide an environment wherein consumers have confidence in the safety of their food supply (Tompkin, 2001).

Changing consumer demands

Consumer demands can be defined as the consumer's requirements to the characteristics of a product, process or service which satisfy the needs of the consumer (Slack *et al.*, 1998). Successful chains in today's competitive agri-food markets operate more market oriented than ever before, with the ultimate goal of responding to changing consumer demands. In their choice of food products consumers are interested in pre-packed, convenience, ready-to-eat, healthy and safe food products (Rabobank, 2002b). Following the recent successive crises in agri-food supply chains, these changing consumer demands coincide with a focus on the reinforcement of quality guarantees at all stages in the agri-food supply chains, from individual farmers to large retailers (Mazé, 2002). Although agri-food products never had such high food quality standards as nowadays, consumers want to know more about the products (including the manner of production) they buy than ever before (Rabobank, 2002b). This places strong demands on gathering, storing, processing and transfer of information between the firms in the chain (Jahn *et al.*, 2004).

Societal demands for Corporate Social Responsibility (CSR)

According to the European Corporate Sustainability Framework (ECSF), CSR can be defined as the inclusion of social and environmental concerns into corporate decision making and business operations as well as the interaction of businesses with stakeholders (Van den Brink and Van der Woerd, 2004). Due to changing attitudes in society, resulting from higher education and increasing wealth, society sets its requirements to firms in agri-food supply chains at a higher level. In the past the provision and security of cheap food was the most important priority for society. Nowadays these requirements are replaced by concerns about the need for sustainable production, not least for production with attention to ethical working practices and ethical trade (Rabobank, 2002b). The various social and environmental (management) standards that currently exist, relate closely to the concepts and disciplines of quality management (Van den Brink and Van der Woerd, 2004). Moreover, many common quality systems do not only take the physical product quality and safety into account, but also focus on organisational quality and environmental, health and labour aspects, e.g. labour circumstances and child labour (Van der Spiegel, 2004).

Willingness to pay for a quality label

Quality management systems can be an effective vehicle to communicate value-related aspects of agri-food products to consumers (Skytte and Blunch, 2001; De Haes *et al.*, 2004; Van den Brink and Van der Woerd, 2004). This is interesting because consumers are usually not willing to pay extra for efforts solely to assure the quality and safety of food. They assume that their food should be safe and should have a high quality. The combination of safety and quality issues with issues consumers are interested in is advantageous for retailers. For example, Dickinson *et al.* (2002) and Hobbs (2003) show that traceability does not deliver much value to consumers, but when it is combined with other characteristics (often so-called credence characteristics) such as animal welfare or environmental friendly production, consumers are willing to pay for that. However, quality and these other characteristics cannot be guaranteed if control systems do not address all stages of the agri-food supply chain (Van Kleef *et al.*, 2006). Through higher revenues obtained by willingness to pay the motivation of firms to integrate their quality management systems in chains might be increased, because firms may see their quality efforts rewarded.

Globalisation of import and export

Globalisation, together with increased international competition, has changed the production, trade and distribution of food products. Nowadays, demand is no longer confined to local or regional supply (Trienekens and Omta, 2002). Nations are becoming increasingly dependent on international traded food products, often at the expense of traditional agricultural commodities (Hooker, 1999). As a result of increasing globalisation food safety problems of one country can easily become the problems of another country. As food may be a transport means for food borne pathogens, globalised food trade may be a mechanism for the spread of food borne pathogens (Motarjemi *et al.*, 2001). Especially developing countries have achieved a growth in the export of non-traditional agricultural speciality products for example, fruits, vegetables, seafood and meats. These products have a great potential for food safety risks. As a result, these commodities are subject to increasing scrutiny and regulations in developed countries as food safety hazards are better understood and more often traced to their sources (Unnevehr, 2000). Therefore, globalisation of food trade has focused the attention of importing firms on strengthening measures to ensure the quality and safety of imported foods. Traditional sampling methods and analysis programs to assure quality of the products are no longer considered as adequate, but nowadays quality assurance should rely on long-term relationships with reliable partners (Hardman *et al.*, 2002; Grievink *et al.*, 2003).

Different quality management regulations/systems

At the moment, a large number of different public and private quality regulations exist. The determination of equivalence of these quality regulations is one of the most important contemporary food safety issues in international trade, because it will help to ensure fair competition among countries in terms of trade (Hathaway, 1999; Motarjemi *et al.*, 2001). Dependent on the country of the buyer, quality requirements will vary among buyer-supplier relationships. For example, British firms require the private quality system BRC, whereas German and French retailers demand QS or IFS for their suppliers (see also Section 2.6). Also public regulations often vary from country to country, because populations around the world differ in terms of their perceptions, values, cultures, religions, lifestyles, needs, motivations and levels of education. In order to successfully conduct trade in food an importing firm must be satisfied that imports meet its legitimate food safety requirements and an exporting firm must judge about the effectiveness of sanitary measures undertaken in the importing country. The development of the SPS Agreement (see Section 2.5) of the World Trade Organisation is a major step in reducing the number of different quality regulations of food quality and food safety management systems (Jukes, 1995).

6.2.2 Drivers from the task environment

Increasing power dependency in the chain

Building relationships in which power is present is highly pertinent to agri-food supply chains. The food industry is becoming more concentrated in all parts of the supply chain, caused by backward vertical integration initiated by powerful retailers and large food firms (Borch *et al.*, 2004). The interest of such chain leaders in safeguarding food safety is strongly related to their legal obligations and to financial and reputational risks in case of food crises (Havinga, 2006). Many of the quality requirements of powerful chain leaders have progressively shifted to process-based controls rather than product inspections (Henson and Loader, 2001). In this regard, the integration of quality management might reduce costs and risks for retailers and inspire confidence in food quality and safety of consumers. Strong chain leaders have set up effective guidelines for managing relationships with suppliers for informing the public adequately with regard to the quality and safety of food. The demands of retailers, often summarised in certification systems, represent the retailers' unique safety and quality requirements (see also Section 2.6).

Chain wide innovation of quality management systems

Most businesses face challenges in maintaining competitiveness and adjusting to changing needs of consumers. One way of gaining competitive advantages is to find new ways of creating added value based on innovative technological developments (Mark-Herbert, 2004). Many of the recently introduced innovative technologies aim at the improvement of quality management (Novoselova *et al.*, 2004). In recent years in literature increasing research attention has been drawn to the make, buy or co-operate decision in innovation. The capability of building inter-organisational network relationships such as buyer-supplier partnerships and strategic alliances is increasingly viewed as the key factor in successful innovations (Pannekoek *et al.*, 2005). It has been stated in previous studies that being part of a supply chain and to be able to effectively exploit information in it has become even more valuable than being able to generate knowledge in the own firm for innovation in quality management (Gambardella, 1992).

6.3 Ranking of the drivers

Sawtooth Software ACA System 5.1. and SPSS 12.0.1 were used for data management and analyses. Table 6.3 shows the means of the relative impacts of drivers on quality management for the three chains as well as for all experts pooled together (the last column). It was very surprising that between the chains hardly any significant differences between the means of the impacts of the drivers on quality management were found. To test these findings, the Kruskal-Wallis tests with significance levels of $p \leq 0.05$ and $p \leq 0.10$ were conducted.

Table 6.3. The means, relative impacts of the different drivers on quality management across the three chains (ranking between brackets).

Drivers	Poultry meat N = 15	Fruit and Vegetables N = 17	Flowers and potted plants N= 15	Total N = 47
Different quality regulations/systems	13.80 (1)	12.92 (2)	11.53 (3)	12.77 (1)
Media attention	12.28 (2)	13.48 (1)	11.83 (2)	12.57 (2)
Increasing power dependency in the chain	9.06 (5)	8.86 (4)	11.86 (1)	9.88 (3)
National legislative demands	10.11 (3)	7.98 (7)	9.67 (4)	9.19 (4)
Changing consumer demands	9.93 (4)	8.44 (5)	8.97 (6)	9.08 (5)
Chain wide innovation of quality management systems	8.42 (6)	9.20 (3)	7.45 (7)	8.39 (6)
Societal demands for corporate social responsibility	6.15* (8)	8.48 (6)	9.75* (5)	8.13 (7)
Willingness to pay for a quality label	8.04 (7)	6.34 (10)	6.89 (8)	7.07 (8)
Globalisation of import	6.12 (9)	7.84 (8)	5.67 (10)	6.59 (9)
Information exchange by ICT	5.95 (10)	6.69 (9)	6.56 (9)	6.51 (10)
Globalisation of export	4.79 (11)	5.13 (11)	5.37 (11)	5.10 (11)
Supra-national legislative demands	5.04 (12)	4.63 (12)	4.47 (12)	4.72 (12)
Total	100	100	100	100
Average consistency (mean of the ACA model fit R^2)	0.69	0.79	0.76	0.75

N represents the number of experts within a group
** difference significant at a $p \leq 0.05$
* difference significant at a $p \leq 0.10$

The drivers in Table 6.3 are listed according to the descending impact on quality management (last column). The impacts of drivers are expressed in percentages, for example, a value of 12.77 in a cell means that a driver counts for an impact of 12.77% of the total impact on quality management of all drivers together. The average consistency is quite high for all the chains involved in this study. Therefore, it can be concluded that experts were highly consistent in answering the questions.

The impact of the drivers on quality management was also linked to the professional background of the experts, see Table 6.4. The ranking of drivers by professional background showed a comparable pattern. Therefore, it can be concluded that the professional background of the experts hardly mattered for the ranking of most of the drivers.

Table 6.4. The means, relative impacts of the different drivers on quality management across professional backgrounds (ranking between brackets).

Drivers	Research N = 17	Business N = 30	Total N = 47
Different quality regulations/systems	13.86 (1)	11.95 (2)	12.77 (1)
Media attention	12.22 (2)	12.83 (1)	12.57 (2)
Increasing power dependency in the chain	8.72 (6)	10.74 (3)	9.88 (3)
National legislative demands	9.44 (4)	9.03 (5)	9.19 (4)
Changing consumer demands	10.40* (3)	8.13* (6)	9.08 (5)
Chain wide innovation of quality management systems	8.96 (5)	7.97 (7)	8.39 (6)
Societal demands for corporate social responsibility	6.35** (9)	9.43**(4)	8.13 (7)
Willingness to pay for a quality label	6.89 (8)	7.17 (8)	7.07 (8)
Globalisation of import	7.26 (7)	6.11 (10)	6.59 (9)
Information exchange by ICT	6.21 (10)	6.73 (9)	06.51 (10)
Globalisation of export	5.92 (11)	4.49 (12)	5.10 (11)
Supra-national legislative demands	3.77 (12)	5.42 (11)	4.72 (12)
Total	100	100	100
Average consistency (mean of the ACA model fit R^2)	0.81	0.71	0.75

N represents the number of experts within a group
** difference significant at a $p \leq 0.05$
* difference significant at a $p \leq 0.10$

In the ranking of the drivers according to the three chains and the professional background the same pattern was visible. From these results it is clear that a stable and robust ranking of drivers from the environment existed[50]. A likely explanation for the low number of differences between the drivers could be that most quality management systems in agri-food supply chains have a comparable design. Quality management systems in all three supply chains included in the present study share activities such as exchange of data on inspection and audit results and evaluation of buyer satisfaction, because these systems are focused on the compliance with norms and requirements. Therefore, the impact of drivers from the business environment might affect especially the structure of the quality management systems instead of the contents.

6.3.1 Highly ranked drivers

The highly ranked drivers are 'different quality regulations/systems' and 'media attention' which both had an impact greater than 10% (see Table 6.3).

---
[50] Hierarchical cluster analysis and the K-means cluster analysis were also carried out, but did not provide new information.

Different quality regulations/systems

The driver 'different quality regulations/systems' received the highest overall rank. Many experts mentioned the increasing work load due to different quality systems. Each system requires its own measures, registrations and information transfers, which have to be made compatible with the original quality management system of the firm. Some experts believed that in the future different quality regulations would increasingly be harmonised, for example, due to intervention of the EU (see Section 2.5.) or large retail organisations, such as the GFSI (see Section 2.6). These organisations have the ability and the power to determine equivalence of quality regulations or systems. Other experts expected that only more different regulations and systems would be developed and a harmonisation of quality regulations would never be achieved. Some experts added that a far-reaching harmonisation of quality regulations between countries would even not be possible because consumers in different countries have different requirements to quality. Interesting was the remark of an expert who stated that harmonisation and differentiation of quality systems would likely occur simultaneously:

> *Existing quality regulations will be more and more harmonised, but the problem is that every moment new quality systems are being developed, which will have to be harmonised over and over again.*
> Senior consultant international trade affairs in the flower and potted plant chain

According to some experts, obstacles for harmonisation of quality regulations were often political or had their origin in trade conflicts. In addition, some experts mentioned that, although some legislative regulations were harmonised in the EU, not all countries apply harmonised regulations in the same way. The Netherlands was often regarded as a country with a stringent execution of quality regulations. It turned out that experts associate the problems of different quality management systems strongly with legislative demands.

Media attention

All experts stated that high chances of negative 'media attention' have a large impact on quality management systems because firms in this case have to prepare more preventive measures in order to anticipate on negative media attention in the future. One firm questioned had already developed a media plan in order to minimise the impact of potential negative media attention. Other experts added that it is never known exactly what the media will pay attention to, making preventive actions for negative media attention almost impossible. One expert suggests that some firms use their recalls even for publicity purposes.

One respondent mentioned that some 'scandals' originate from a lack of knowledge of the media. For example, in the past there was negative media coverage of the sales of Dutch and other European poultry meat in Western Africa. Some organisations suggested that these products were of low quality and were therefore dumped with the help of subsidies. However, these sales concerned a common practice on the world market for poultry meat: chicken legs and fowls were difficult to sell in Western Europe and considered as delicacy in Western Africa. Although in particular the poultry meat chain has faced a lot of negative media attention in the past, this driver has not received a higher score in this chain compared to the two other chains.

6.3.2 Medium ranked drivers

The medium ranked drivers are 'increasing power dependency in the chain', 'national legislative demands', 'changing consumer demands', 'chain wide innovation of quality management systems' , and 'societal demands for corporate social responsibility'. All these drivers have an impact between 8% and 10%.

Increasing power dependency in the chain

In agri-food supply chains power is generally skewed in favour of large retailers due to their large buying power. Their suppliers (traders, processors and primary producers) are the dependent firms (Hingley, 2005). Many of the private quality management systems as discussed in Chapter 2 are designed by (associations) of retailers, see Table 2.8. According to a majority of experts, increasing power dependency in the chain has an important impact on quality management systems, because a strong chain leader with a clear interest in quality can apply its power to urge less powerful suppliers to comply with increasingly stringent quality requirements. Quality requirements of British retailers are in particular stringent as one expert stated:

> *If these detailed requirements from British retailers would decrease; quality management would be less complex. Each retailer wants its own certifications and audits, with only slight differences which make quality management very costly.*
> Quality manager of a fruit and vegetable trading firm

However, other experts expected that increased power dependency would have less impact on their quality management systems. According to them the judicious use of power by a chain leader may add to the development of more uniform quality measurements. It is interesting that especially in the flower and potted plant sector, increasing power dependency in the chain receives a very high rank. This high score might be the result of the planned merger between the two main flower auctions in The Netherlands (Flora Holland, 2007).

National legislative demands

The driver 'national legislative demands' was defined as the level of interaction of the government with firms on quality issues. A majority of the experts indicated that increasing interaction of the government with the industry would have an important impact on quality management systems, because it often resulted in more legislative demands with regard to quality. These legislative demands, which often change, have to be included in their quality management systems. However, other experts indicated that national legislative demands were not important to them at all because the quality requirements of their buyers were much more stringent. It is striking that the importance of national legislative demands with regard to quality in the poultry meat chain and fruit and vegetable chain was not regarded as significantly higher than in the flower and potted plant chain. Food safety in particular has been placed high on the regulatory agendas and has resulted in many new quality legislative demands. A possible explanation can be that the flower and potted plant chain has also faced many legislative demands, but especially in the field of environmental and labour issues.

Changing consumer demands

For the driver 'changing consumer demands', there was a weak significant difference ($p \leq 0.10$) between experts from business and research. A likely explanation is the perception of practical problems on quality issues in trade. Compared to experts from business, experts from research focus more on the need of information gathering, storing and processing. They think that invisible characteristics of products and processes that satisfy buyers have to be verified. They especially worry about how to assure these characteristics when suppliers from developing countries are involved in the chains. These suppliers often lack adequate quality management systems. Experts from business have a less complicated view by stating that all their suppliers have to comply with their quality requirements as one expert clearly stated:

> *I give them the specifications and they have to comply with these specifications, because otherwise I will switch to other suppliers.*
>
> Importer of poultry meat

According to most experts changing consumer demands will increasingly have an impact on quality management systems in their chains. In the past, the price was especially important for consumers, but nowadays consumers in developed countries do not face financial limitations any more. As a result, consumers increasingly select the products they buy according to their personal demands such as quality, healthiness and convenience. This development is present in each chain involved in this study (Rabobank, 2002a, b)

Chain wide innovation of quality management systems

According to a majority of the experts, participation in strongly integrated chains places strong quality demands on the firms. Each firm in a closely integrated chain has to give insight in its quality procedures. A 'chain wide innovation of quality management systems' is desirable because if improved systems are implemented in an individual firm, the whole chain may have benefits. For example, problems often arise at the beginning of the chain (with feed suppliers or primary producers) and or the end of the chain (with the consumers). If firms have a strong notion that the whole chain is responsible for quality management, quality requirements will be implemented very easily even if these quality requirements are very high. Moreover, experts added that in collaboration environments the wheel is not reinvented and bargaining problems are prevented. However, the start of collaboration in innovation in the field of quality management may be difficult, but it is inevitable as one respondent stated:

> *Chain-wide innovations for quality management systems require a lot of transparency between firms, such as sharing information. Nowadays this openness is still scarce; however, it is the success for the future.*
>
> Respondent from a flower auction

Societal demands for corporate social responsibility

For the driver 'societal demands for corporate social responsibility' (CSR) no consensus exists. In the flower and potted plant chain CSR is regarded to have much more impact on quality management than in the poultry meat chain ($p \leq 0.10$). The reason might be that in the flower and potted plant chain CSR plays a more important role in quality management systems compared with the other two chains. For example, MPS, an important quality management system in the flower and potted plant chain (see also Section 2.10), is aimed at the protection of the environment. Furthermore, it has a module which takes care of labour practices. In addition in the poultry meat chain and the fruit and vegetable chain quality is especially aimed at food safety issues, and to a less extent to CSR issues, such as environmental protection. Professional background also seems to be important for societal demands for CSR, because according to experts from business CSR had much more impact on quality management than according to the experts from research. However, this was the result of the low number of experts from research in the flower and potted plant chain.

6.3.3 Lowly ranked drivers

Lowly ranked drivers are 'willingness to pay for a quality label', 'globalisation of import', 'information exchange by ICT', 'globalisation of export' and 'supra-national legislative demands'. These drivers have an impact smaller than 8% of the total.

Willingness to pay for a quality label

The relative low score for the 'willingness to pay for a quality label' is likely influenced by the fact that most experts thought that consumers are not willing to pay extra for a quality label. A number of experts stated that many firms do not get any premium price paid for products produced under a certain quality label, because consumers regard safe food as a threshold requirement. Experts further indicate that producers have to guarantee the claims of the label consumers were paying for in their quality management systems.

Globalisation of import

The driver 'globalisation of import' received a remarkable low score for its impact on quality management. Experts provided probable explanations. They mentioned that suppliers from developing countries also had to comply with quality requirements of the EU. Other experts reported that they sourced their products from big commercial farms with European management in these countries. According to them these firms had comparable quality performances. The three chains included in the present study differ (although not in a strict statistical way) with regard to the import of products. The fruit and vegetable chain imports large quantities of products, whereas the poultry meat chain and the flower and potted plant chain are mainly supplied by domestic producers. Therefore, globalisation of import receives the highest rank in the fruit and vegetable chain.

Information exchange by ICT

It is remarkable that the driver 'information exchange by ICT' did not get a higher overall rank, regarding the emphasis that has been placed on the use of ICT in Supply Chain Management literature. Most experts emphasised the standardisation of quality data due to information exchange by ICT which enables them to handle the increasing amount of quality data in their quality management systems. However, some experts mentioned that before implementing such systems, clear appointments in the chain have to be made. Furthermore, the willingness of firms to share information is much more important as the technical possibilities. These experts regard ICT only as a solution, a supportive tool for the organisation of quality management as a respondent stated:

> Information exchange by ICT is supportive for quality management in the chain. However, first quality management has to be organised in the chain and after that ICT is a means to implement it.
>
> Quality manager of a poultry slaughterhouse

Globalisation of export

The striking low score for 'globalisation of export' is the result of the fact that firms have to comply with the quality legislation of the country where the products were produced. Therefore, the destination of the products (developing or developed countries) has less impact on quality management. However, some experts mentioned the huge strictness of idiosyncratic quality regulations in some countries, especially with regard to phytosanitary requirements. They perceived that there was often no scientific justification as one expert stated:

> These requirements differ from time to time and it seems that requirements are adapted randomly.
>
> Representative from an interest organisation in the flower and potted plant chain

Sometimes experts expect that these requirements are used to protect the home market. Others stated that some countries, for example Japan and Australia, had such stringent requirements, because they were afraid of importing diseases.

Supra-national legislative demands

The driver 'supra-national legislative demands' received the lowest overall rank. As was already stated, some countries had very specific and stringent quality requirements. With regard to other countries experts did not expect many problems for exporting to outside the EU, because EU quality regulations were already among the most stringent in the world. For importing products to the EU, experts stated that these products also had to comply with the quality legislation from the EU. Other experts added that they were already performing beyond legislative quality compliance, because many private quality management systems go beyond the European legislative quality requirements.

6.4 Concluding remarks

In this chapter, 47 experts ranked drivers from the general and task environment of the poultry meat chain, the fruit and vegetable chain and the flower and potted plant chain which have an impact on quality management. It turns out that ACA is very useful in Supply Chain Management studies. The findings indicate that a clear ranking of these drivers exists, over different chains and professional backgrounds (experts from business and research). An exception is the driver *'societal demands for corporate social responsibility'*. This driver is significantly more important in the flower and potted plant chain than in the poultry meat and fruit and vegetable chain. The likely explanation is that in the other two chains food safety issues have dominated the contents of quality management systems, whereas in the flower and potted plant chain other issues, especially environmental and labour issue have been important. In the next phase of this study (presented in the next chapter), the most important drivers from the general business environment found in this chapter are included in the survey. The most important drivers were:
- Different quality regulations/systems
- (National) legislative demands
- Chances of negative media attention
- Changing consumer demands
- Societal demands for corporate social responsibility.

Regarding the drivers 'different quality regulations' and 'national legislative demands', these drivers were strongly related to each other and combined in one more general driver, *'legislative demands'*. The formulation of a broader driver *'legislative demands'* enables the supranational legislative demands to be taken into account too. The most important drivers from the task environment, increasing power dependency and chain wide innovations in quality management systems are largely covered by the construct 'integration of quality management systems' (see

Chapter 7). The most important drivers found in this chapter are included in the survey in order to investigate their impact on quality management. The outcomes of the survey are presented in the next chapter.

Chapter 7. Survey results

In this chapter characteristics of the study samples and the results of the statistical analyses are reported. For data management and analysis the statistical software packages SPSS 12.0.1 and Lisrel 8.72 were used. This chapter starts with a discussion of the response to the questionnaire in Section 7.1. In Section 7.2 some characteristics of the six samples included in this study are described. Section 7.3 discusses the analysis of the response, non-response bias and informant selection. In Section 7.4 two measurement- and two structural models are developed based on the research model, one for the supplier side and one for the buyer side of the focal firm. Subsequently, the generalisability of both the measurement and the structural models across groups is investigated. Section 7.5 assesses the validity and reliability of the constructs. Section 7.6 discusses the outcomes of the estimated structural models. The chapter ends with some concluding remarks in Section 7.7.

7.1 Response

For collecting the data 3,312 questionnaires were sent in the period September-December 2005 of which 585 useable questionnaires were returned, a response rate of 19% (18% useable). This response rate is satisfying in the light of the fact that no reminder questionnaire was sent. Table 7.1 shows the distribution of the response across the six groups of firms involved in this study.

Looking at the absolute numbers of questionnaires returned it can be concluded that the objective to receive approximately one hundred questionnaires for each group[51] in order to be able to use adequately statistical methods (Hair *et al.*, 1998) has been achieved except for the poultry meat processors. For the primary producers this level has been exceeded, especially for the growers of fruit and vegetables.

Although the response rate was not high for poultry meat processors, data on the number of employees and the yearly turnover showed that many large firms in this chain had returned the questionnaire. Taking a turnover of 5 million Euros per year as a cut-off value between small and large firms, a response rate of 35% was achieved for this group[52]. For fruit and vegetables, 24 traders and/or processors with a turnover of 25 million Euros per year or more returned the questionnaire. A turnover of 25 million Euros a year is regarded as the turnover

---

[51] For adequate regression for each predictor at least five observations should be availabe. The maximum number of predictor variables during the regression analyses was fourteen. This would imply a minimum of 70 questionnaires per group of firms. However, it is preferred to have more observations per predictor, so the most peferred group size is approximately 100 or larger.

[52] According to a spokesman of the Product Boards of Livestock, Meat and Eggs in the Netherlands there are approximately seventeen big slaughterhouses and forty big cutters (Ms. Ariënne Visser, personal phone call, December 2005). She stated that previous research in this sub-group had obtained lower response rates.

Table 7.1. Response rate across the six groups of firms.

	Poultry meat		Fruits and vegetables		Flowers and potted plants		Total
	Farmers[c]	Traders/ processors[d]	Growers[e]	Traders/ processors	Growers[e]	Traders/ processors	
Total sample mailed	599	313	600	600	600	600	3,312
Non eligible firms[a]	8	4	1	12	3	8	36
Incomplete	5	1	6	4	11	9	36
Useable	116	34	151	98	102	84	585
Response rate (%)[b]	20	11	26	17	19	16	19

[a]Non eligible firms are duplicate addresses, liquidated firms and firms who have changed their activities.
[b]Response rate = (total number of returned questionnaires)/ (total sample mailed - non eligible firms), useable response rate: (total number of returned questionnaires-incomplete questionnaires)/ (total sample mailed - non eligible firms).
[c]All firms with more than 35.000 chickens (Product Boards for Livestock, Meat and Eggs).
[d]All processors present in The Netherlands (Product Boards for Livestock, Meat and Eggs).
[e]Growers with more than 10.000 m² greenhouse.

dividing small and large traders on the domestic market (see also Table 2.5 in Section 2.3). If it is assumed that all these 24 traders and processors are active on the domestic market, a response rate of 67% has been achieved in this group. In the flower and potted plant chain 58 traders exist with a turnover of more than 10 million Euros per year. From this group at least 38 traders have returned the questionnaire implying a response rate of 65% for that group. For primary producers the selection of larger firms had already been made before sending the questionnaire, because much more information about size for these groups of firms was available at the Product Boards. Data from large firms is advantageous, because these firms cover a much larger part of the total market.

Besides the useable questionnaires, 36 incomplete and blank questionnaires were returned. Some respondents indicated that the questions were too difficult, too scientific, too abstract or too general. Also 36 non eligible questionnaires were returned. Sometimes firms did not exist any longer or had changed their activities, for example, some poultry farmers did no longer fatten up chickens, but had switched to egg production. There were also firms which had very specialised and uncommon products, such as aquarium plants. Of course these firms were not included in the analyses.

7.2 Characteristics of the study samples

This section describes some characteristics of the study populations, starting with some *general* characteristics which were measured in all firms. After that some *specific* characteristics of primary producers and traders and/or processors are discussed.

7.2.1 General characteristics

In Table 7.2 the number of employees (own and hired personnel in full time equivalents) and the yearly turnover are shown as indicators for size. In general, primary producers are much smaller than traders and/or processors and among the primary producers, poultry farmers represent firms with the smallest size. Regarding the standard deviations it becomes clear that within the groups considerable differences exist with regard to the number of employees and turnover. However, this could be expected given the skewed distribution between the number of firms and market shares in all chains (see Chapter 2).

The revenue growth compared with the competitors (see Table 7.2) is for the six groups around the four, the centre of the measurement scale[53]. This means that the firms included in the study have a revenue growth that is comparable with their main competitors. Therefore, respondents seem to represent firms with average revenue growths in their sectors. Traders from the poultry meat chain and the fruit and vegetable chain perceive that their revenues grow just a little bit slower compared to their main competitors. Poultry farmers and traders of flowers and potted plants think that their revenues grow somewhat faster than the revenues of their main competitors.

Looking at the expected change in turnover it becomes clear that most firms are positive about the future. It is remarkable that although the poultry traders and/or processors and the fruit and vegetables traders and/or processors indicate that their competitors would realise a better revenue growth, they are quite positive about their expected annual growth of turnover. However, one should notice that the data concern an estimation of expected growth of turnover and not a realised turnover at the time the questionnaire was filled out[54]. For each chain, traders achieve a higher expected growth of turnover compared to the primary producers. The limited possibilities for expansion in some parts of The Netherlands, for example, due to light emission or spatial planning are possible reasons for lower expected growth of the turnover for growers.

---

[53] Growth of revenues compared with main competitors was measured on a quasi perceptual seven point Likert scale on which 1 means much slower; 4 comparable and 7 much faster.

[54] This reason was often mentioned by respondents for not providing an answer on this question. The questionnaire offers space to add comments on all questionnaire related topics.

Table 7.2. The average number of employees (own and hired personnel; fte), average yearly turnover, revenue growth compared to main competitors and expected annual growth of turnover (standard deviation between brackets).

Kind of firms	Average number of employees (fte)	Average yearly turnover (million Euros)	Revenue growth compared to competitors	Expected annual growth of turnover (in %)
Poultry meat				
Farmers	1.3 (0.6)	0.8 (0.4)	4.4 (1.1)	2.8 (7.0)
Traders/Processors	63.1 (69.0)	13.9 (18.7)	3.8 (1.6)	6.1 (6.0)
Fruits and vegetables				
Growers	9.5 (8.2)	1.3 (1.0)	4.0 (1.3)	3.2 (5.3)
Traders/Processors	21.4 (20.6)	21.9 (27.8)	3.8 (1.4)	4.2 (6.0)
Flowers and potted plants				
Growers	15.1 (21.5)	2.2 (2.8)	4.2 (1.3)	3.8 (4.6)
Traders/Processors	26.8 (29.5)	15.8 (17.8)	4.5 (1.1)	5.1 (5.3)

For poultry farmers the import of cheap poultry meat from Brazil and Thailand, and the recent Aviary Influenza ('bird flu') could be explanations for the relatively low growth expectations.

Due to increased attention for quality management, many firms have implemented quality management systems and or have employed a quality manager. In the investigated chains, quite general quality systems exist which are often prerequisites for participation in trade (see for description of quality management systems Chapter 2). Integraal Keten Beheer (in English: Integrated Chain Control) serves as such a system in the poultry chain in The Netherlands in which for example, 90-95% of the poultry farmers participate (Van Horne et al., 2006). Table 7.3 shows the number of quality management systems, employment of a quality manager and presence of a chain leader on quality.

Primary producers almost all possess a quality system and often this system is a kind of standardised system, such as IKB, Eurep-GAP or MPS. Firms that have no quality management system often work according to certain hygiene codes, such as in the fruit and vegetable chain the 'Basis zorgsysteem' of the Greenery (a big fruit and vegetable trader and auction in The Netherlands) whose requirements are comparable to Eurep-GAP. For traders of flowers and potted plants, no standard quality system exists. Regarding the participation of firms in 'standard' quality management systems, the samples seem to represent their populations rather well.

Table 7.3. Number of quality management systems, presence of a quality manager and quality chain leader across the different kinds of firms in the three chains.

Kind of firm	Number of quality management systems (%)				Standard quality system[a] (%)	Quality manager (%)	Quality chain leader (%)
	None	I	2 or more	Total			
Poultry meat							
Farmers	6	70	24	100%	93	4	53
Traders/Processors	12	39	49	100%	88	56	39
Fruits and vegetables							
Growers	I	38	61	100%	93	19	66
Traders/Processors	24	36	40	100%	52	63	62
Flowers and potted plants							
Growers	3	62	35	100%	89	27	53
Traders/Processors	71	16	13	100%	-	46	30

[a]For the poultry meat farmers: IKB; For the poultry processors/traders: IKB or HACCP; For fruit and vegetable growers: Eurep-GAP; For fruit and vegetable traders: HACCP; For flower and potted plant growers: MPS A, B or C; For flower and potted plant traders: none

Traders who have more quality systems often use ISO 9000 as a second quality system. Moreover, for traders the variation in the number of quality management systems implemented is somewhat greater than for the primary producers. An explanation can be that sometimes no widely used general quality systems exist for traders, such as for the traders of flowers and potted plants, explaining the high number of traders with no quality management system at all in that sample. Another reason could be that quality systems are not obligatory for traders who do not carry out any physical treatment on the product itself, like some traders in the fruit and vegetable chain. However, since January 2006[55] all traders in this sector have to work according to HACCP or to a hygiene-code which could be regarded as a sector wide translation of HACCP, which are often set up by Product Boards.

The high score of both the fruit and vegetable growers and fruit and vegetable traders for the presence of a chain leader for quality, compared to the other two sectors is remarkable. A chain leader is defined as a partner in the chain who is able to enforce its quality requirement on other partners in the chain. Several explanations can be given for this difference. Compared to traders in flowers and potted plants, fruit and vegetable traders more often deliver to

---

[55] This questionnaire was sent in 2005.

large retailers who are able to enforce quality requirements in the chain. The big size of the poultry processors compared to other groups of firms could explain that they do not so often experience a chain leader. Due to the limited number of poultry processors, combined with their large size, retailers have fewer possibilities to switch from one poultry processor to another. Poultry farmers indicate that they do not encounter a chain leader on quality often. A possible explanation is that many slaughterhouses accept the system IKB as a proof of good quality management, which almost all poultry farmers have in practice.

Not surprisingly, traders which are on average larger than primary producers, more often employ a quality manager than primary producers. For small firms it is not profitable and often not necessary to employ a quality manager, because the amount of quality related tasks is much smaller and less diverse than for big firms. In small firms quality management is often one of the tasks of the owner and for building or adapting the quality management systems small firms often hire external expertise.

7.2.2 Specific characteristics of the primary producers

For primary producers some specific characteristics were measured such as the age of the respondent and whether or not a successor was present at the firm. Figure 7.1 presents the distribution of the age of primary producers included in this study[56] and compares it with data about the age of the eldest owner of the firm according to LEI (Landbouw Economisch Instituut; in English: Agricultural Economics Research Institute). The categories of LEI are somewhat broader than the categories used in this study. For example, the category domestic animals includes all farmers with all kind of cattle, besides poultry.

Not surprising regarding the workable age in each group 90% or more of the respondents are between thirty and sixty-five years old and the category forty through forty-nine is the largest category in each chain. The respondents included in the study seem to be rather young compared to the data provided by the LEI[57]. A possible explanation can be that LEI measures the age of the eldest owner. This is important because a common practice in The Netherlands is that many farms are owned by a partnership which often consists of two or more family members, for example, father and son. The group respondents of the flower and potted plant chain seem to be the youngest compared to the other two chains. A possible reason that relatively young primary producers are included in this study can be that they are more open to research or are more aware of the need for quality related research.

For the older respondents it was investigated whether or not a successor was present (Figure 7.2). Many older respondents have no successor for their firms, especially for growers of flowers

---

[56] In the questionnaire for traders and processors, the questionnaire did not ask the age of the respondents, because it was expected that age would only have an effect on the business processes of primary producers.

[57] The LEI data about age do not contain the confidence intervals needed to calculate whether respondents in the present study are significantly younger than those included in the LEI-data.

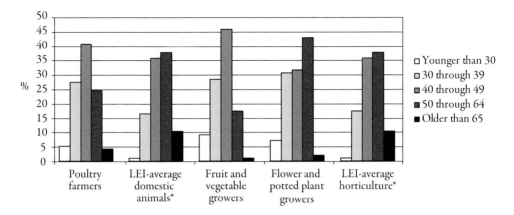

Figure 7.1. Age of the respondents compared with LEI-averages.
Source: CBS (2006)

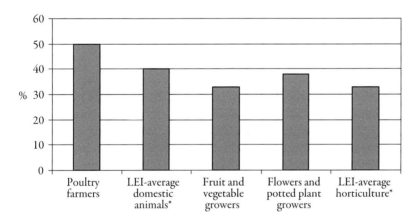

Figure 7.2. Percentages of firms where the respondent was 50 years or older and a successor was present.
Source: CBS (2006).

and potted plants. Regarding the data of LEI, the phenomenon of having no successor is a common problem for older primary producers, because only 40% of the primary producers with domestic animals older than fifty have a successor and only 33% of the primary producers from horticulture older than fifty have a successor (Berkhout and Van Bruchem, 2006).

Poultry farmers and growers of flowers and potted plants older than 50 years included in this study seem more often to have a successor compared to the LEI-data. This is not the case for growers of fruit and vegetables. In general the larger the firm, the more likely the farmer has a successor.

7.2.3 Specific characteristics of the traders

Also for traders some specific characteristics were measured, such as the percentage of sales that was generated on the domestic and international markets and the number of buyers and suppliers. In Figure 7.3 the percentage sales generated on the international market is shown for traders.

As can be concluded for Figure 7.3 most of the traders and/or processors of poultry meat realise their sales on the domestic market. The traders of fruit and vegetables and flowers and potted plants are much more internationally oriented. Moreover, one should notice that these percentages do not say anything about the volume of products sold on the domestic or international market. A closer analysis revealed that big traders especially sell their products on the international markets. Therefore, an analysis based on the volumes of products could be quite different; however, this information was not available. Traders that were active in foreign countries often sell their products in countries such as Germany, the United Kingdom, Belgium and France. Traders in flowers and potted plants additionally sell their products to a less extent in Russia, Scandinavia, Italy and Switzerland. These findings are in accordance with literature and the chain descriptions in Chapter 2.

In order to get an insight into how a supply chain (see Section 3.1) looks for the traders/processors the percentages of firms with less than 20, between 20 and 50, between 50 and 150 and more than 150 suppliers/buyers are presented in Figure 7.4. Most traders and/or processors in each chain have less than 20 suppliers, whereas poultry processors most often have the largest number of buyers, likely due to their large size compared with traders and/or processors in other chains. Slaughterhouses in the group poultry traders and/or processors may buy from a high number of poultry farmers, causing the high score in the category between 50 and 150 suppliers. A remarkable fact is that traders of flowers and potted plants most often have between 20 and 50 buyers, but no specific reason was available for this phenomenon.

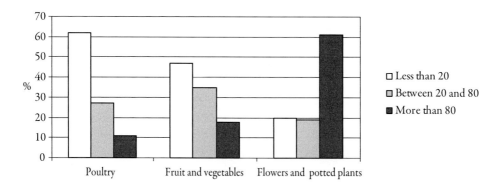

Figure 7.3. Percentage trader sales generated on the international market.

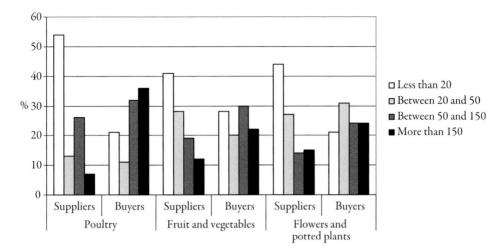

Figure 7.4. Percentage of traders or processors with less than 20, between 20 and 50, between 50 and 150 and more than 150 suppliers/buyers for groups of traders/processors.

7.3 Analysis of response, non-response bias and informant selection

Insight in the representativeness of the study sample is provided by the analysis of the response. Table 7.1 has already shown that the total response was quite balanced among the different groups of firms, except for the poultry processors and traders, who were a minor group of only 6% of the total number of firms included in the study. If this group is not taken into account, the response percentages of the different groups in the total sample are between 14% and 26%. This means that none of the groups of firms is dominant in this study sample. Also when the total sample is segmented in sub-samples, for example, only primary producers, no specific group of firms is dominant.

The representativeness of the samples is this study was tested by carrying out a non-response bias analysis using the extrapolation method proposed by Armstrong and Overton (1977). In this method the assumption is made that respondents who returned the questionnaire less readily are more like non-respondents in their answering pattern. In many studies 'less readily' is defined as answering the questionnaire after sending a reminder. However, in this study no reminder was sent and therefore, the time series approach was used to make a distinction between early and fast respondents, based on the time of return. This approach has the advantage that the possibility of a bias introduced by sending the reminder itself is eliminated. The questionnaires were entered in chronological sequence in the data management system and for each group of firms the last 25% of the respondents were regarded as late respondents. A t-test was used to discover differences of the scores on the constructs for early and late respondents. No significant differences ($p \leq 0.05$) were found in each sub-group. Therefore, non-response is not regarded as a problematic bias in this study and the samples could be regarded as representative for each group of firms.

The selection of appropriate respondents is important in order to get reliable answers and a high quality of the response. For the primary producers in this study the questionnaire was focused on the owner of the firm. Regarding the size of these firms, it is most likely that this person could provide the data. Indeed it turned out that for most primary producers by far the questionnaire was filled out by the owner, ranging from 80% to 95% across groups. For the traders and/or processors the questionnaire was addressed to the persons responsible for quality management in the firms. It turned out that most times this was the owner, a member of the board or a quality manager, ranging from 76% to 80% for the traders and/or processors. Regarding these results it can be assumed that the respondents have sufficient knowledge to provide the relevant data.

7.4 Analysis of generalisability

In studies in which data form different independent samples are analysed, like in this study, a serious limitation is often the assumption that the set of items and the number of underlying constructs has to be the same across all samples (Baumgartner and Steenkamp, 1998). If between groups of firms meaningful comparisons are to be made, the *measurement* model has to be invariant for all the firms (Steenkamp and Baumgartner, 1998). According to Steenkamp and Baumgartner (1998) if the purpose is to compare structural models in a nomological net, metric invariance and factor[58] variance invariance are required. Metric invariance means that the factor loadings on the constructs do not differ significantly across the groups. Factor variance invariances means that the variances of the constructs do not differ significantly across the groups (see Section 5.2.4). To further improve the results the covariance invariance was also tested. More stringent constrained models are always preferable, because they further strengthen the conclusions. Covariance invariance implies that co-variances between the constructs in the groups of firms do not differ significantly and that the paths in the structural model for each group are not likely to be significantly different. In order to test this last statement (whether or not the paths in the structural model differ from each other), the invariances of the paths in the *structural* models were also tested. The separate assessment of the measurement - and the structural models is known as the two-step approach (Anderson and Gerbing, 1988).

7.4.1 Analysis of the generalisability of the measurement models

Because the research model is applied to the supplier and buyer side of the focal firm, for each perspective a separate model was developed. All questions regarding the supplier and buyer model were filled in by one respondent of the focal firm. Figure 7.5 shows which constructs are included in each model.

---

[58] For the word factor, also the word construct could be used.

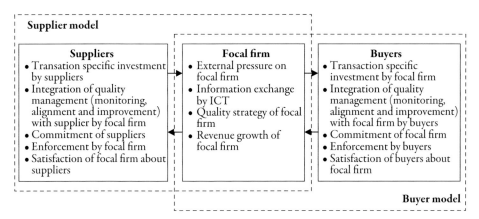

Figure 7.5. Constructs in the supplier and the buyer model.

Constructs that are specific for the focal firm, for example external pressure, are included in both models. For measuring constructs that focus on the relationship between the focal firm and the supplier or the buyer, the same kind of questions were used making the models comparable, although the perspective from the focal firm was different (see for exact formulation of the questions also Appendix 4). Both supplier and buyer model were tested on metric invariance and factor (co)variance invariance by comparing four multi-group models. In each multi-group model six groups (based on the kind of firms included in the study poultry farmers, poultry processors, fruit and vegetable growers, fruit and vegetable traders, flower and potted plant growers and flower and potted plant traders) were compared. In Table 7.4 the scores on the recommended fit indices are shown for the different supplier models[59].

In the *first* model, invariance constraints were imposed on the factor loadings, factor variances and factor co variances. This means that the factor loadings, factor variances and factor co-variances for each construct are the same in each group. This model shows a good fit on the selected fit indices $\chi^2/df = 1.46$ (should be lower than 2 or 3)[60]; RMSEA= 0.07 (should be lower than 0.08); CAIC= 3971.97 (no specific threshold, especially used for model comparisons); NNFI = 0.93 (should be higher than 0.90); CFI = 0.93 (should be higher than 0.90). The meaning of these fit indices can be found in Box 5.2. This model will serve as a base-line model to which other models will be compared on the basis of selected fit indices.

---

[59] The single item factors external pressure, information exchange by ICT and revenue growth of the focal firm are not included in the measurement model. The researcher gives these indicators an estimate, which is the same for all groups, because it turned out that the standard deviations of the single indicators were very close to each other. When the single item indicators are tested for factor variance invariance and factor covariance invariance no problems were encountered for these indicators.

[60] Large values of $\chi2$ to the degrees of freedom indicate that observed and estimated matrices differ considerably (Kemp, 1999). However, this ratio may be higher in case of model complexity and large sample size (Laros and Steenkamp, 2004).

Table 7.4. Model comparisons for the supplier model.

Model	$\chi^2 (\Delta\chi^2)*$	df (Δdf)	p	CAIC	RMSEA	NNFI	CFI
Factor loadings and factor (co) variances constrained	2564.69	1757	<0.01	3971.97	0.07	0.93	0.93
(Co) variances constrained, factor loadings relaxed	2429.18 (135.51)	1672 (85)	<0.01	4456.23	0.07	0.93	0.93
Factor loadings constrained (co) variances relaxed	2267.49 (297.20)	1577 (180)	<0.01	4170.84	0.07	0.93	0.94
Factor loadings and (co) variances relaxed	2162.77 (396.92)	1492 (265)	<0.01	5502.313	0.07	0.93	0.94

*$\Delta\chi^2$ and Δdf with baseline model (first model).
Critical value of $\Delta\chi^2$ for 80 degree of freedom $\approx$ 107, ($p \le 0.05$), critical value of $\Delta\chi^2$ for 180 degrees of freedom $\approx$ 213, ($p \le 0.05$), critical value of $\Delta\chi^2$ for 265 degrees of freedom $\approx$ 303.

For each model the change of Chi-Square and change of the number of degrees of freedom compared with the baseline model are shown (Chi-Square Difference Test). However, the Chi-Square Different Test is sensitive to large sample sizes and model complexity. As a result Chi-Square Difference Test of complex and large sample models become extremely easy significant. Therefore, Steenkamp and Baumgartner (1998) recommend to use the NNFI, CFI, RMSEA and the CAIC for model comparisons. Higher values of the NNFI and CFI and lower values of the RMSEA and CAIC indicate better models. RMSEA, NNFI and CAIC (of which the CAIC is the most powerful) are particularly useful, because they were found to be the most effective indices in distinguishing between correctly and incorrectly specified models. These fit indices take into account both goodness of fit and model parsimony[61] by imposing a penalty on fitting additional parameters.

In the *second* model factor variance and factor co-variances were constrained whereas, no constraints on the factor loadings were imposed. This means in practice that factor loadings on each factor could vary across the groups, but factor variance and factor co-variances were held the same across groups. As could be expected the Chi-Square Difference Test was significantly lower compared with the baseline model. However, compared to this model, the baseline model shows better values for the CAIC and the NNFI, whereas the RMSEA and the CFI hardly change between the models. Therefore, it is concluded that this model does not fit better as the baseline model and holding factor loadings fixed across the groups is not a problem, which implies that metric invariance is supported.

---

[61] Parsimony could be regarded as the trade off between simplicity of the models and the predictiveness of different models.

In the *third* model, factor loadings are constrained among the groups, but the factor variances and factor co-variances are not constrained[62]. This means that the variances of the factors themselves and the co-variances between the factors could vary across groups, but the factor loadings for each factor are the same across groups. Again, the baseline model shows a better fit on the CAIC and the NNFI, whereas the other model has a marginal better fit on the RMSEA and the CFI. So it is not demonstrated that the alternative model fits better than the baseline model, although the Chi-Square Difference Test is significant which could be expected regarding the model complexity and large sample size. This implies that factor variance invariance and factor covariance invariance are supported in this study.

In the *fourth* model, as a last check no constraints were imposed on the factor loadings, factor variances and factor co-variances, so they could all vary across groups. This model shows a worse fit on the CAIC and NNFI compared to the baseline model, whereas the CFI and RMSEA slightly improve.

For the buyer model the same procedure was repeated and the results are shown in Table 7.5. Also in the buyer model in which metric and or factor (co) variance invariance were imposed does not show a worse fit as models in which metric variance and or factor (co) variance variances were allowed. Moreover, the model with factor invariance and (co) variance invariances constraints imposed shows a good fit: $\Delta\chi^2/df = 1.50$; RMSEA= 0.07; CAIC= 4026.26; NNFI = 0.92; CFI = 0.92.

Table 7.5. Model comparisons for the buyer model.

Model	χ^2 $(\Delta\chi^2)*$	df (Δdf)	p	CAIC	RMSEA	NNFI	CFI
Factor loadings and factor (co) variances constrained	2633.57	1757	<0.01	4026.26	0.07	0.92	0.92
(Co) variances constrained, factor loadings relaxed	2448.44 (185.13)	1672 (85)	<0.01	4460.91	0.07	0.92	0.93
Factor loadings constrained (co) variances relaxed	2283.00 (350.57)	1577 (180)	<0.01	4988.17	0.07	0.92	0.93
Factor loadings and (co) variances relaxed	2133.51 (500.06)	1496 (265)	<0.01	5458.46	0.07	0.92	0.93

$*\Delta\chi^2$ and Δdf with baseline model (first model).
Critical value of $\Delta\chi^2$ for 80 degree of freedom $\approx$ 107, ($p \leq 0.05$), critical value of $\Delta\chi^2$ for 180 degrees of freedom $\approx$ 213, ($p \leq 0.05$), critical value of $\Delta\chi^2$ for 265 degrees of freedom $\approx$ 303.

---

[62] Also a model was tested in which only the factor variances were constrained. This model did not result in a better model fit as the baseline model.

From these tests it can be concluded that for both the supplier model and the buyer model metric and factor (co) variance invariance is supported and a good fit to the data is retained. This means that comparisons of the structural models of the groups of firms are possible and no wrong conclusions are made due to variances in the measurement model. No detailed group analyses are needed with regard to the *measurement* model and the data of the groups can be pooled together (Laros and Steenkamp, 2004).

Having pooled the data of the groups together, constructs were investigated in a separate Confirmatory Factor Analyses (CFA) and are assessed on the magnitude of the factor loadings, t-values, and fit indices. (CFA is explained in detail in Appendix 1, see Box A1.1). The most important feature of CFA is that it offers the possibility to test if prior notion about which variables load on which factors is consistent with the patterns in the data (Lattin *et al.*, 2003). During this partial analysis no serious problems were encountered for the separate constructs. After that a CFA was carried out on the total measurement model for the supplier model and the buyer model. This approach can be regarded as the ultimate test for the validation of the constructs, because in one overall CFA it is possible for errors between all indicators to freely correlate[63]. Because all the data were pooled together other fit indices also become relevant for the assessment of the model fit, such as the NFI, GFI and AGFI, see Table 7.6. Box 5.2 in Section 5.2.4 contains more information about these fit indices. The CAIC is less relevant, because this indicator is especially important for comparing models. To assess and to evaluate how well the specified model accounts for the data multiple indices should be examined (Anderson and Gerbing, 1988; Steenkamp and Van Trijp, 1991; Jöreskog and Sörbom, 1996).

The χ^2 is significant, but with large sample sizes, significant values can be obtained even though there are only trivial discrepancies between a model and the data and the same holds for more complex models (Anderson and Gerbing, 1988; Laros and Steenkamp, 2004). Therefore, the ratio between χ^2 and the number of degrees of freedom is used. Both measurement models

Table 7.6. Fit indices for the assessment of the fit for the supplier and buyer measurement model.

Model	χ^2	df	p	χ^2/df	RMSEA	NFI	NNFI	CFI	GFI	AGFI
Supplier	599.69	257	<0.01	2.33	0.05	0.97	0.98	0.98	0.92	0.90
Buyer	693.34	257	<0.01	2.70	0.06	0.96	0.97	0.98	0.91	0.88

---

[63]Moreover, by using one big CFA for the total model the problem of saturated models (resulting in perfect model fits) is also avoided. In that case no assessment of the fit of the individual constructs can be made, because the number of degrees of freedom is zero. Degrees of freedom are used to calculate the fit indices; so without them it becomes impossible to assess the fit of the model. Researchers usually solve this problem by estimating multiple constructs simultaneously.

Integration and self regulation of quality management

show acceptable fits, which means that the measurement models give a good representation of the underlying covariance matrices. For the supplier model, the ratio between the Chi square and the degrees of freedom is 2.33 which is satisfying, especially considering the large sample size (Laros and Steenkamp, 2004). Moreover, the RMSEA is far below the threshold level of 0.08 and the other fit indices also have higher levels to the recommended level of 0.90. Also the AGFI, which is sensitive to non normality, is far above its threshold level of 0.80. The measurement model for the buyers shows an almost comparable fit as the supplier model, but scores just a little bit lower on all fit indices, see Table 7.6. The complexity of the model is not a problem and the buyer model is also be acceptable. The reported values of the standardised factor loadings, errors, t-values and R^2's in Section 7.5.2 for testing the validity and reliability of the reflective constructs arc based on the CFAs of the total measurement models of the supplier and buyer model presented here.

7.4.2 Analysis of the generalisability of the structural models

After having obtained validated and reliable measurements, the hypotheses in the research model were tested. Hypothesis testing could be regarded as the second step in the two step approach for structural equation modelling as proposed by Anderson and Gerbing (1988). The last hypothesis proposed in Chapter 4 predicting that integration of quality management is positively related to performance was split up into a number of sub-hypotheses (e.g. the direct impact of integration on the performance indicators, buyer satisfaction and revenue growth and these relationships mediated by the dimensions of self regulation, commitment and enforcement). Furthermore, it was also tested whether or not increasing buyer satisfaction leads to a higher revenue growth. For each group of firms a separate structural model was composed in order to investigate if one overall model could be used for the description of the relationships across groups, by testing the invariances of the paths in the structural models. In the model for testing the path invariances the metric and (co) variance invariances constraints were imposed, which was allowed according to the tests conducted on the measurement model in the previous section. For the evaluation of the path invariances, the same fit indices as for testing the measurement model were used.

Firstly, a constrained model was tested with equality constraints for all the paths coefficients across all the groups. This means that the standardised coefficients of the paths (structural relationships) were kept constant across the groups. *Secondly*, equality constraints for each path were set free one at a time and the fit indices were compared with the fit indices of the fully constraint model (Ahire and Dreyfus, 2000). This means that the standardised coefficient of one path at a time could vary across the groups. *Thirdly*, a test was performed in which the totally constrained model was compared with the model in which all paths were set free. The results of these tests for the supplier model are depicted in Table 7.7 The estimated baseline models show a good fit, $\chi^2/df = 1.42$; RMSEA = 0.069; NNFI = 0.92; and CFI = 0.92. The Hs in Table 7.7 refers to the hypothesis see Table 7.29. that was set free that time.

Chapter 7

Based on the Chi-Square Difference Test, some paths are significant different across the groups[64] (paths with $\Delta\chi^2 > 11$), but with regard to the other fit indices no support is provided, because again, the most important indicator for distinguishing correctly and incorrectly specified models, the CAIC, has the lowest value for the fully constrained model. Also the differences for the RMSEA, NNFI and CFI between the fully constrained model and models with one path set free each time or all paths set free at once were extremely small, indicating that path invariance was not a problem. (see Table 7.7).

The baseline buyer model also shows a good fit, $\chi^2/df = 1.54$, RMSEA $= 0.08$, NNFI $= 0.90$ and CFI $= 0.90$, see Table 7.8 The values for the NNFI and CFI are just on the recommended level of 0.90, but this is not a problem regarding the large sample size and model complexity.

Table 7.7. Outcomes for testing the path invariances of the supplier model.

Constrained H	χ^2	p	$\Delta\chi^2$*	df	Δdf*	RMSEA	CAIC	NNFI	CFI
All	3194.36	< .01	-	2245	-	0.07	4587.05	0.92	0.92
All except:									
H_1	3178.12	< 0.01	16.24	2240	5	0.07	4607.27	0.92	0.92
H_2	3179.86	< 0.01	14.50	2240	5	0.07	4609.01	0.92	0.92
H_3	3191.87	< 0.01	2.49	2240	5	0.07	4594.29	0.92	0.92
H_4	3165.10	< 0.01	29.26	2240	5	0.07	4621.02	0.92	0.92
H_{5a}*	3179.36	< 0.01	15.00	2240	5	0.07	4608.51	0.92	0.92
H_{5b}*	3184.37	< 0.01	9.99	2240	5	0.07	4613.52	0.92	0.92
H_6	3182.39	< 0.01	11.97	2240	5	0.07	4611.54	0.92	0.92
H_{7a}	3175.51	< 0.01	18.85	2240	5	0.07	4604.66	0.92	0.92
H_{7b}	3184.64	< 0.01	9.72	2240	5	0.07	4613.79	0.92	0.92
H_{7c}	3182.13	< 0.01	12.23	2240	5	0.07	4611.28	0.92	0.92
H_{7d}	3190.75	< 0.01	3.61	2240	5	0.07	4619.90	0.92	0.92
H_{7e}	3169.24	< 0.01	25.12	2240	5	0.07	4604.66	0.92	0.92
H_{7f}	3187.30	< 0.01	7.06	2240	5	0.07	4616.45	0.92	0.92
H_{7g}	3193.27	< 0.01	1.09	2240	5	0.07	4615.22	0.92	0.92
All free	3049.91	< 0.01	144.45	2175	70	0.07	4953.01	0.92	0.92

*5a Commitment → integration of quality management.
5b Integration of quality management → Commitment.
$\Delta\chi^2$ and Δdf compared with baseline model (first model).

---

[64] Critical value of $\Delta\chi^2$ for 5 degree of freedom $= 11.07$, (p $\leq$ 0.05), critical value of $\Delta\chi^2$ for 65 degrees of freedom ≈ 91, (p $\leq$ 0.05).

162   Integration and self regulation of quality management

In the buyer model, according to the Chi-Square Difference Test, some paths are significantly different across the groups, but regarding the other fit indices no support is being provided, because again, the CAIC has the lowest value for the fully constrained model and the RMSEA, NNFI, CFI are marginally changed. Also the differences for the RMSEA, NNFI and CFI between the fully constrained model and models with all paths set free at once were extremely small, indicating that path invariance was not a problem in the buyer model either (see Table 7.8).

The outcomes of path invariances are not so surprising given the fact that the co-variances between the factors were also invariant. Regarding the outcomes of testing the metric invariance, factor (co) variance invariance and path invariance it can be concluded that the data of the six groups can be pooled all together (Laros and Steenkamp, 2004) both with regard to the measurement model as well as the structural model. The structural models in which all data were pooled together show a good fit to the data. These fit indices are reported in Figure 7.13 and Figure 7.14.

Table 7.8. Outcomes for testing the path invariances of the buyer model.

Constrained H	χ^2*	p	$\Delta\chi^2$*	df	Δdf*	RMSEA	CAIC	NNFI	CFI
All (baseline)	3450.33	< 0.01		2245		0.08	4828.43	0.90	0.90
All except:									
H_1	3440.52	< 0.01	9.81	2240	5	0.08	4855.10	0.90	0.90
H_2	3445.82	< 0.01	4.51	2240	5	0.08	4860.82	0.90	0.90
H_3	3440.53	< 0.01	9.80	2240	5	0.08	4855.09	0.90	0.90
H_4	3449.87	< 0.01	0.46	2240	5	0.08	4864.43	0.90	0.90
H_{5a}*	3445.88	< 0.01	4.45	2240	5	0.08	4860.45	0.90	0.90
H_{5b}*	3447.12	< 0.01	3.21	2240	5	0.08	4861.68	0.90	0.90
H_6	3429.66	< 0.01	20.67	2240	5	0.08	4844.23	0.90	0.90
H_{7a}	3443.73	< 0.01	6.60	2240	5	0.08	4858.30	0.90	0.90
H_{7b}	3446.50	< 0.01	3.83	2240	5	0.08	4861.07	0.90	0.90
H_{7c}	3431.87	< 0.01	18.46	2240	5	0.08	4846.43	0.90	0.90
H_{7d}	3447.41	< 0.01	2.92	2240	5	0.08	4861.97	0.90	0.90
H_{7e}	3440.22	< 0.01	10.11	2240	5	0.08	4854.79	0.90	0.90
H_{7f}	3440.86	< 0.01	9.47	2240	5	0.08	4855.42	0.90	0.90
H_{7g}	3446.26	< 0.01	4.07	2240	5	0.08	4860.82	0.90	0.90
All free	3342.87	< 0.01	107.46	2175	70	0.08	5231.39	0.90	0.90

*5a Commitment → integration of quality management.
5b Integration of quality management → Commitment.
$\Delta\chi^2$ and Δdf compared with baseline model (first model).

7.5 Analysis of the validity and reliability of the constructs

In this study, both reflective and formative constructs are used. The nature of formative or causal constructs is opposite to those of reflective constructs. For formative constructs items can be viewed as causing the variable rather than that the items are caused by the variable. Formative constructs are composed of items which directly represent the operational definition and are regarded as explanatory combinations of items. Due to the fact that the items of a formative construct determine the construct and that omitting one item is omitting a part of the construct (Bollen and Lennox, 1991), methods used to asses the validity and reliability of reflective constructs are not suitable for formative constructs. The difference between formative and reflective constructs is discussed extensively in Appendix 1. First the validation and reliability of formative constructs are discussed below.

7.5.1 Formative constructs

Formative constructs are validated by an assessment of the content validity (*domain*, do they measure what they intend to measure, and *history*, how did they perform in previous research), nomological validity (how well it is related to other theoretically related constructs) and multicollinearity (correlation between variables of the indicators), which are discussed in Appendix 1. This study contains two formative constructs, external pressure and information exchange by ICT.

Content validity

The construct of external pressure on the focal firm is strongly based on the extensive literature study presented in Section 6.2. and interviews with experts involved in the EU-concerted action Global Food Network. Furthermore, from the Adaptive Conjoint Analysis (ACA) presented in Chapter 6 the most important drivers from the business environment were identified. Combining these two approaches resulted in items that capture the relevant drivers from the business environment assuring the content validity of the construct. The questions used for the operationalisation of the drivers were derived and combined from previous studies which were related to the motivation for implementation of management systems for corporate social responsibility and environmental care in which they perform well (Klassen and Angell, 1998; De la Cruz Déniz Déniz and Suárez, 2005). For each driver the set of questions were comparable. The individual drivers from the business environment are transferred into a perceived environmental pressure index (see Section 5.2.2).

In the questionnaire one question was included about information exchange by ICT. Initially this question was regarded as a control variable, because integrated information exchange by ICT did not receive a very high rank during the ACA, see Chapter 6. However, during the analysis, the scores on this question turned out to be highly significantly related with the integration of quality management in both the supplier and the buyer model and also in the analyses on the sub-groups. In addition, during the in-depth interviews about the 'best practices' for quality management, many respondents stressed the importance of information exchange by ICT. Therefore, it was decided to include information exchange by ICT as an external predictor for the integration of quality management. This question was adapted from Chen and Paulray (2004).

A final test of the reliability of the formative constructs external pressure and information exchange by ICT was achieved during the pre-test of the survey. Potential respondents and experts involved in the pre-test of the questionnaire perceived the questions in a similar way which means that content validity and reliability of the formative constructs has been achieved (Churchill, 1999).

Nomological validity

Nomological validity of the formative constructs was achieved by finding statistical significant relationships with the constructs they were expected to be related to. In Section 7.6 the structural models are discussed, in which the strengths and significances of the relationships can be found.

Multicollinearity

Multicollinearity of the set of items that compose formative constructs is also checked by examining the strength of the correlation between the items composing the construct (Diamantopoulos and Winklhofer, 2001). The coefficients do not suggest any obvious problem of item multicollinearity that would preclude their use. Correlations were below 0.80 which is often regarded as an indication for non-multicollinearity (Malhotra *et al.*, 1999).

7.5.2 Reflective constructs

The reliability and validity of the reflective constructs, quality strategy, transaction specific investment, integration of quality management, commitment, enforcement, buyer satisfaction and revenue growth are assessed in this section according to the procedures described in Appendix 1. For these constructs, content, nomological and convergent validity are assessed as well as the reliability.

Content and nomological validity

Content validity was assessed in the same way as for formative constructs, by adapting validated scales from previous studies, interviews with experts and thorough pre-testing of the questionnaire. The reflective constructs were acceptable in terms of history and domain. Nomological validity was also achieved because in the structural models (see Section 7.6) many significant relationships were found with other reflective and formative constructs and the constructs were behaving as expected.

Convergent validity

For the assessment of the convergent validity and further evaluation of discriminant validity, the item-total correlations, factor loadings from both the explorative factor analysis and confirmatory factor analysis and variance explained were investigated. The objective of Explorative Factor Analysis (EFA) is to identify common factors and explain the relationships to the observed data (Lattin *et al.*, 2003). An important note is that the derived structure of the EFA is data driven and can be different from the structure that could be expected from theory. Confirmatory factor analysis (CFA) offers the possibility to test if prior notions about which variables load on which factors is consistent with the patterns in the data (Lattin *et al.*, 2003). Furthermore, the reliability of the constructs by evaluating the Cronbach's α, composite reliability and variance extracted is judged. In Appendix 4 the questions corresponding to abbreviations used in the tables and figures are reported.

Quality strategy of the focal firm

Quality strategy of the focal firm was measured with three items. Table 7.9 shows that item-total correlations and factor loadings were all above the threshold levels of respectively 0.50 and 0.60 (Hair *et al.*, 1998; Kemp, 1999). Because the quality strategy was measured for the individual firms, the same construct is used in both the buyer and supplier model. Therefore, no separate outcomes for the supplier and buyer model are presented. Including the construct quality strategy of the focal firm in the supplier or buyer model did not result in different factor loadings in the CFA.

The explained variance exceeds with 72.9% the recommended level of 60% far. The outcomes of the CFA show (Figure 7.6.) that all λ 's have higher loadings than the required threshold level of 0.60 and are highly significant. The arrows in the figure represent the graphical way of relating items to constructs in Lisrel.

Table 7.9. Item-total correlation and factor loadings (EFA) for quality strategy of the focal firm.

Items	Both models	
	---	---
	Item-total correlations	Factor loadings
QSl	0.61	0.82
QS2	0.72	0.89
QS3	0.65	0.85

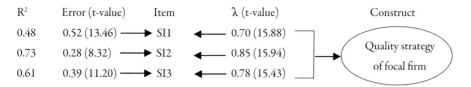

R²	Error (t-value)	Item	λ (t-value)	Construct
0.48	0.52 (13.46) ⟶	SI1	⟵ 0.70 (15.88)	
0.73	0.28 (8.32) ⟶	SI2	⟵ 0.85 (15.94)	Quality strategy of focal firm
0.61	0.39 (11.20) ⟶	SI3	⟵ 0.78 (15.43)	

Figure 7.6. Confirmatory factor analysis for the construct quality strategy of focal firm.

No problems were encountered with regard to the reliability coefficients of the construct quality strategy of the focal firm, see Table 7.10. Cronbach's α and composite reliability are much higher than 0.70. These indicators measure the extent to which indicators 'share' in their measurement of a construct. High reliability means that they all are measuring the same construct (Hair *et al.*, 1998). The level of variance extracted by the construct quality strategy of the focal firm is above the threshold level of 0.50. Variance extracted is the amount of 'shared' or common variance among the items for a construct. Higher values represent a greater degree of shared representation of the indicators with the construct (Hair *et al.*, 1998). From these analyses it can be concluded that the construct quality strategy of the focal firm is both valid and reliable.

Table 7.10. Reliability coefficients for quality strategy of focal firm.

Construct	Both models		
	Cronbach's α	Composite reliability	Variance extracted
Quality strategy of the focal firm	0.81	0.82	0.61

Transaction specific investments (TSIs)

The construct TSIs was measured by four items. The item-total correlations and the factor loadings depicted in Table 7.11 are all above the threshold levels. The explained variances for the supplier and buyer model were satisfied with respectively 78.2% and 72.6%.

The CFA also shows a good support of the data of the TSIs construct (Figure 7.7). All factor loadings were highly significant and the lowest standardised factor loading (0.75) was much higher than the threshold level of 0.60.

Table 7.11. Item-total correlation and factor loadings (EFA) for TSIs.

By supplier (supplier model)			By focal firm (buyer model)		
Items	Item-total correlations	Factor loadings	Items	Item-total correlations	Factor loadings
TsiS1	0.77	0.82	TsiC1	0.70	0.83
TsiS2	0.84	0.92	TsiC2	0.76	0.87
TsiS3	0.83	0.91	TsiC3	0.77	0.88
TsiS4	0.72	0.84	TsiC4	0.69	0.82

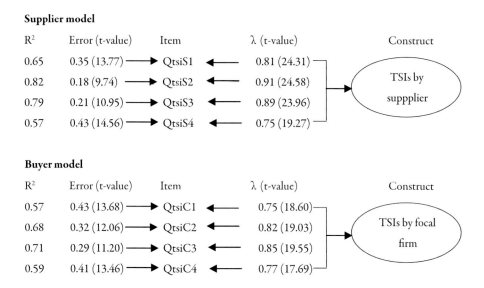

Figure 7.7. Confirmatory factor analysis for TSIs: supplier model above and buyer model below.

The construct TSIs achieves high values on the Cronbach's α, composite reliability and variance extracted (Table 7.12). Therefore, it can be concluded that the construct TSIs is valid and reliable in both models.

Table 7.12. Reliability coefficient for TSIs.

Construct	By supplier (supplier model)			By focal firm (buyer model)		
	Cronbach's α	Composite reliability	Variance extracted	Cronbach's α	Composite reliability	Variance extracted
TSIs	0.91	0.91	0.71	0.87	0.88	0.64

Integration of quality management

The second order construct integration of quality management was measured in three dimensions, monitoring (three items), alignment (three items) and improvement of integration of quality management (four items). Table 7.13 shows that the item-total correlations between the items were higher than 0.50 in both models. All factor loadings are far above 0.60 indicating good convergent validity of the construct. The explained variance was good: 79.1% for the supplier model and 83.6 % for the buyer model for monitoring, 80.0% and

Table 7.13. Item-total correlation and factor loadings (EFA) for integration of monitoring, alignment and improvement in quality management.

By supplier (supplier model)			By focal firm (buyer model)		
Items	Item-total correlations	Factor loadings	Items	Item-total correlations	Factor loadings
Monitoring					
MonS1	0.72	0.88	MonC1	0.61	0.87
MonS2	0.78	0.91	Monc2	0.68	0.89
MonS3	0.74	0.89	Monc3	0.67	0.89
Alignment					
AlignS1	0.76	0.90	AlignC1	0.73	0.84
AlignS2	0.79	0.91	AlignC2	0.79	0.80
AlignS3	0.73	0.88	AlignC3	0.78	0.83
Improvement					
ImproS1	0.70	0.84	ImproC1	0.70	0.83
ImproS2	0.71	0.85	ImproC2	0.73	0.82
ImproS3	0.76	0.87	ImproC3	0.71	0.87
ImproS4	0.63	0.78	ImproC4	0.71	0.84

80.4% for respectively the supplier and buyer model for alignment and 69.7% and 72.2% for improvement in respectively the supplier and buyer model.

In the CFA monitoring, alignment and improvement of quality management were combined to estimate the second order construct integration of quality management. Figure 7.8 and Figure 7.9 shows the outcomes for respectively the supplier and the buyer model.

For both models, the factor loadings of the first order constructs and the loadings of the first order constructs on the second order constructs are good, except the loading of monitoring

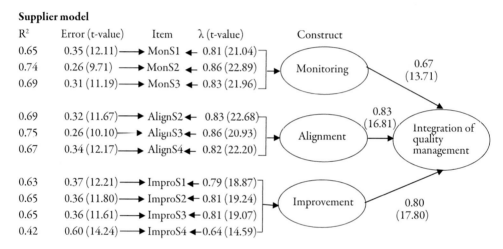

Figure 7.8. Confirmatory factor analysis for integration of quality management: supplier model.

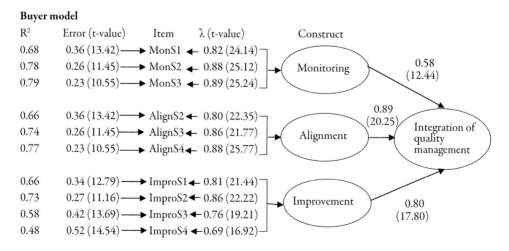

Figure 7.9. Confirmatory factor analysis for integration of quality management: buyer model.

falls just below the threshold level of 0.60 in the buyer model. Table 7.14 presents the reliability coefficients of the first order constructs monitoring, alignment and improvement. The reliability coefficients for the second order construct integration of quality management are also shown. All the obtained reliability indicators are far above the threshold levels. From these analyses it is clear that both first and second order constructs are valid and reliable in both models.

Table 7.14. Reliability coefficients for integration of quality management.

Construct	By supplier (supplier model)			By focal firm (buyer model)		
	Cronbach's α	Composite reliability	Variance extracted	Cronbach's α	Composite reliability	Variance extracted
Monitoring	0.87	0.87	0.70	0.90	0.90	0.75
Alignment	0.88	0.87	0.70	0.88	0.90	0.72
Improvement	0.85	0.85	0.58	0.87	0.86	0.61
Integration of quality management	0.89	0.81	0.59	0.91	0.86	0.68

Commitment

Commitment was measured with three items. In Table 7.15 the item-total correlations are all above the recommended level of 0.50. The explained variance is good: 83.1% for the supplier model and 74.5% for buyer model. Factor loadings of the EFA are all far above the threshold level of 0.60.

Table 7.15. Item-total correlation and factor loadings (EFA) for commitment.

By supplier (supplier model)			By focal firm (buyer model)		
Items	Item-total correlations	Factor loadings	Items	Item-total correlations	Factor loadings
ComS1	0.80	0.91	ComC1	0.66	0.85
ComS2	0.83	0.93	ComC2	0.67	0.85
ComS3	0.77	0.89	ComC3	0.73	0.89

Figure 7.10 shows the satisfactory results of the CFA. All factor loadings are above the threshold levels. The reliability coefficients in Table 7.16 show a good reliability for commitment for both models. All reliability indicators exceed the threshold levels.

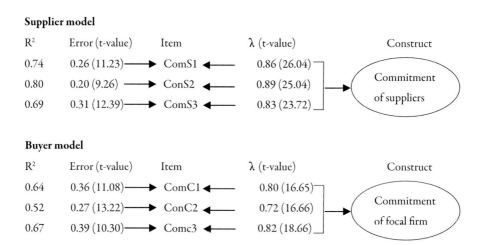

Supplier model

R²	Error (t-value)	Item	λ (t-value)	Construct
0.74	0.26 (11.23) → ComS1 ←		0.86 (26.04)	
0.80	0.20 (9.26) → ConS2 ←		0.89 (25.04)	Commitment of suppliers
0.69	0.31 (12.39) → ComS3 ←		0.83 (23.72)	

Buyer model

R²	Error (t-value)	Item	λ (t-value)	Construct
0.64	0.36 (11.08) → ComC1 ←		0.80 (16.65)	
0.52	0.27 (13.22) → ConC2 ←		0.72 (16.66)	Commitment of focal firm
0.67	0.39 (10.30) → Comc3 ←		0.82 (18.66)	

Figure 7.10. Confirmatory factor analysis for the construct commitment: supplier model above and buyer model below.

Table 7.16. Reliability coefficients for commitment.

	By supplier (supplier model)			By focal firm (buyer model)		
Construct	Cronbach's α	Composite reliability	Variance extracted	Cronbach's α	Composite reliability	Variance extracted
Commitment	0.90	0.90	0.74	0.83	0.82	0.61

Enforcement

Table 7.17 displays the results for the construct enforcement, which was measured with two items. In the supplier model the construct performs well, the item-total correlations are above the 0.50 and the factor loadings exceed 0.60. In the buyer model the item-total correlations are just below the level of 0.50. For the factor loadings no problems were encountered. The explained variance is for the supplier model 80.6% and for the buyer model 74.3%, which is far above the threshold level of 60%.

Table 7.17. Item-total correlation and factor loadings (EFA) for enforcement.

By supplier (supplier model)			By focal firm (buyer model)		
Items	**Item-total correlations**	**Factor loadings**	**Items**	**Item-total correlations**	**Factor loadings**
EnfS1	0.61	0.90	EnfC1	0.49	0.86
EnfS2	0.61	0.90	EnfC2	0.49	0.86

The results of the CFA are depicted in Figure 7.11. All factor loadings exceed the threshold level of 0.60, except the last item in the buyer model which just falls below this level. However, this loading was highly significant.

The reliability indicators for enforcement in the supplier model were highly satisfying. For the buyer model, the Cronbach's α and the composite reliability were just on the threshold levels (Table 7.18). The variance extracted was above the threshold level in the buyer model and therefore the construct is not regarded as problematic.

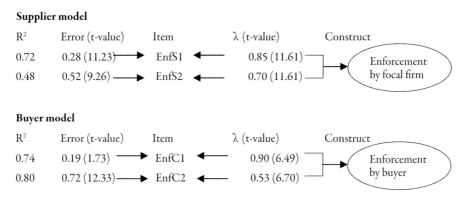

Figure 7.11. Confirmatory factor analysis for the construct enforcement: supplier model above and buyer model below.

Table 7.18. Reliability coefficients for enforcement.

Construct	By focal firm (Supplier model)			By buyers (Buyer model)		
	Cronbach's α	Composite reliability	Variance extracted	Cronbach's α	Composite reliability	Variance extracted
Enforcement	0.76	0.82	0.61	0.65	0.69	0.55

Buyer satisfaction

The construct buyer satisfaction was measured by three items. The total item correlations for all the items in both models are higher than 0.50 and factor loadings are higher than 0.60 (see Table 7.19). The explained variance was good; 77.0% for the supplier model and 77.9% for the buyer model.

Furthermore, CFA indicates (Figure 7.12) a good fit of the data for the construct buyer satisfaction. All λs are highly significant and exceed the threshold level of 0.60.

Values of the reliability indicators are reported in Table 7.20 and all coefficients indicate a high level of reliability of the construct buyer satisfaction in both models, because they are higher than the threshold levels (0.70 for the Cronbach's α and the composite reliability and 0.50 for the variance extracted). Therefore, it can be concluded that buyer satisfaction was a valid and reliable construct.

Table 7.19. Item-total correlation and factor loadings (EFA) for buyer satisfaction.

Satisfaction of focal firm about suppliers (Supplier model)			Satisfaction of buyer about focal firm (Buyer model)		
Items	Item-total correlations	Factor loadings	Items	Item-total correlations	Factor loadings
SatS1	0.77	0.89	SatC1	0.74	0.91
SatS2	0.80	0.92	SatC2	0.80	0.93
SatS3	0.63	0.82	SatC3	0.62	0.81

Integration and self regulation of quality management

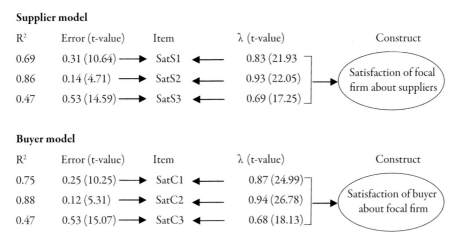

Figure 7.12. Confirmatory factor analysis for buyer satisfaction: supplier model above and buyer model below.

Table 7.20. Reliability coefficient for buyer satisfaction.

Construct	Supplier model			Buyer model		
	Cronbach's α	Composite reliability	Variance extracted	Cronbach's α	Composite reliability	Variance extracted
Buyer satisfaction	0.85	0.86	0.67	0.84	0.87	0.70

Revenue growth of the focal firm

For measuring the revenue growth of the focal firm a quasi perceptual scale was used, in which the respondents could indicate the revenue growth compared to their main competitors. Other financial performance indicators such as the turn-over growth per year and the total turnover per year resulted in high levels of non response rates. Therefore, these scales were only used for descriptive purposes. This phenomenon is quite common in survey research and has lead to researchers asking the same questions in an indirect manner (Ketokivi and Schroeder, 2004). Also during the pre-test of the questionnaire people indicated that the quasi perceptual scale was much more useful than directly asking for financial performance indicators. Therefore, the quasi perceptual scale was considered as the most reliable indicator for measuring and comparing the revenue growth of the focal firm. By using this scale general influences on the revenue growth in a specific group are filtered out, because revenue growth was compared with other firms in the same group. Therefore, valid comparisons between firms from different sectors are possible. Because this reflective construct is measured by a single item, the omnibus of tests carried out on the other reflective scales is left behind.

Discriminant validity

Discriminant validity assesses the extent to which a construct and its indicators differ from other constructs and their indicators. Discriminant validity was assessed by three methods.

Firstly, discriminant validity of a construct is established when the Cronbach's α are larger than the averaged interscale correlations (Ghisellla *et al.*, 1981; Ahire and Dreyfus, 2000; Kaynak, 2003). *Secondly,* if the percentage of variance extracted by the indicators of a construct is consistently greater than the average interscale correlation of the construct, discriminant validity of the construct with respect to all other constructs is achieved (Ahire and Dreyfus, 2000). *Thirdly*, more conservative, discriminant validity is further strengthened if none of the individual interscale correlations of a construct is greater than its Cronbach's α (Fornell and Larcker, 1981). Table 7.21 shows the results of the first two methods for the supplier model. Both criteria show that discriminant validity was achieved in the supplier model, because Cronbach α's are larger than the averaged interscale correlations and also the variance extracted is larger than the squared inter construct correlations for each construct. In Table 7.21, the results for the third method are shown, which show that none of the constructs had an interscale correlations greater than its Cronbach's α, further strengthening discriminant validity.

Table 7.21. Discriminant checks for the supplier model.

	Cronbach's α	Variance Explained (VE)	Average interscale correlation (AVISC)	Cronbach's α - AVISC	VE - AVISC^2
Quality strategy of focal firm	0.81	0.61	0.40	0.41	0.45
TSIs of supplier	0.91	0.71	0.39	0.62	0.56
Integration of quality management	0.85	0.67	0.31	0.54	0.57
Monitoring	0.87	0.70	0.36	0.52	0.57
Alignment	0.87	0.70	0.43	0.44	0.52
Improvement	0.85	0.58	0.44	0.41	0.39
Commitment of supplier	0.90	0.74	0.40	0.50	0.58
Enforcement by focal firm	0.76	0.61	0.41	0.35	0.44
Satisfaction of focal firm about supplier	0.89	0.59	0.46	0.43	0.38

Table 7.22 presents the outcomes for the buyer model which show that discriminant validity is also achieved for this model. Also none of the individual interscale scale correlations of a construct (reported in Table 7.24) were greater than the Cronbach's α. Regarding the discriminant validity checks, it can be concluded that the constructs are clearly distinct from each other. From all the analyses provided above it can be concluded that the models fit well to the data and the constructs are valid, reliable and distinctive from each other for the total model as well as the sub-group models.

Table 7.22. Discriminant checks for the buyer model.

	Cronbach's α	Variance Explained (VE)	Average interscale correlation (AVISC)	Cronbach's α - AVISC	VE - AVISC²
Quality strategy of focal firm	0.81	0.61	0.29	0.52	0.53
TSIs by focal firm	0.87	0.64	0.38	0.49	0.50
Integration of quality management	0.91	0.68	0.40	0.51	0.42
Monitoring	0.90	0.75	0.33	0.57	0.64
Alignment	0.88	0.72	0.42	0.46	0.54
Improvement	0.87	0.61	0.43	0.44	0.43
Commitment of focal firm	0.83	0.61	0.32	0.49	0.51
Enforcement by buyer	0.65	0.55	0.25	0.40	0.49
Buyer satisfaction about focal firm	0.87	0.70	0.30	0.57	0.61

7.6 Analysis of the structural models

Before the structural models are presented, some baseline statistics such as the means, standard deviations of the constructs and the bi-variate Pearson correlations between constructs are presented for both models (Table 7.23 and Table 7.24)

Some remarks should be made regarding the Tables 7.23 and 7.24:
- Correlations higher than 0.08 are significant on a 5% significance level and correlations higher than 0.11 are significant on a 1% significance level. All correlations are positive in the supplier model. The same holds for the buyer model, except the non-significant correlation between enforcement by buyers and revenue growth of the focal firm.
- The high correlations between the constructs monitoring, alignment and improvement should not be taken into account for the assessment of discriminant validity, because these

Table 7.23. Means, standard deviations and correlations of the constructs in the supplier model.

	Mean	Standard deviation	External pressure on focal firm	Quality strategy of the focal firm	TSIs by suppliers	Information exchange by ICT	Integration of quality management with supplier by focal firm	Monitoring	Alignment	Improvement	Commitment of suppliers	Enforcement by focal firm	Satisfaction of focal firm about suppliers	Revenue growth of focal firm
External pressure on focal firm	53.49	20.53	na											
Quality strategy of the focal firm	5.11	1.24	**.39**	.81										
TSIs by suppliers	3.75	1.66	**.40**	**.35**	.91									
Information exchange by ICT	3.74	1.78	**.22**	**.32**	*.17*	na								
Integration of quality management with supplier by focal firm	4.29	1.27	**.52**	**.49**	**.55**	**.36**	.85							
Monitoring	3.35	1.67	**.41**	**.38**	**.49**	**.40**	**.76**	.87						
Alignment	4.21	1.60	**.43**	**.42**	**.40**	**.22**	**.82**	**.42**	.87					
Improvement	5.14	1.41	**.44**	**.39**	**.46**	**.29**	**.87**	**.47**	**.60**	.85				
Commitment of suppliers	5.33	1.35	**.23**	**.31**	**.38**	*.14*	**.48**	**.25**	**.46**	**.43**	.90			
Enforcement by focal firm	4.30	1.62	**.27**	**.24**	**.35**	**.18**	**.45**	**.28**	**.43**	**.39**	**.46**	.76		
Satisfaction of focal firm about suppliers	5.19	1.13	**.19**	**.25**	**.32**	*.07*	**.34**	**.20**	**.29**	**.33**	**.50**	**.26**	.89	
Revenue growth of focal firm	4.12	1.30	*.08*	**.20**	*.11*	**.18**	*.17*	*.12*	*.15*	*.14*	*.10*	*.06*	*.08*	na

Cronbach's α at the diagonal if applicable (na = not applicable).
Bold correlations are significant at a 1% level (two tailed).
Italic correlations are significant at 5% (two tailed).
Correlations that are neither bold nor italic are non-significant (two tailed).
Correlations that are underlined are correlations of first order constructs on their second order construct. These correlations should not be taken into account for the discriminant validity.

Table 7.24. Means, standard deviations and correlations of the constructs included in the buyer model.

	Mean	Standard deviation	External pressure on focal firm	Quality strategy of the focal firm	TSIs by focal firm	Information exchange by ICT	Integration of quality management with focal firm by buyers	Monitoring	Alignment	Improvement	Commitment of focal firm	Enforcement by buyer	Buyer satisfaction about supplier	Revenue growth of focal firm
External pressure on focal firm	53.49	20.53	na											
Quality strategy of the focal firm	5.11	1.24	.39	.81										
TSIs by focal firm	4.47	1.57	.50	.44	.87									
Information exchange by ICT	3.74	1.78	.22	.32	.33	na								
Integration of quality management with focal firm by buyers	4.41	1.35	.49	.34	.58	.31	.91							
Monitoring	3.68	1.94	.39	.32	.57	.33	.73	.90						
Alignment	4.98	1.50	.41	.29	.46	.23	.88	.43	.88					
Improvement	4.32	1.49	.44	.27	.46	.25	.90	.48	.74	.87				
Commitment of focal firm	5.92	0.97	.16	.31	.19	.13	.37	.12	.38	.39	.83			
Enforcement by buyer	4.80	1.45	.23	.11	.29	.12	.37	.25	.33	.34	.24	.65		
Buyer satisfaction about supplier	5.89	0.94	.18	.32	.26	.14	.32	.17	.29	.35	.58	.22	.87	
Revenue growth of focal firm	4.12	1.30	.08	.20	.12	.18	.11	.07	.10	.10	.09	-.01	.14	na

Cronbach's α at the diagonal if applicable.

Bold correlations are significant at a 1% level (two tailed).

Italic correlations are significant at 5% (two tailed).

Correlations that are neither bold nor italic are non-significant (two tailed).

Correlations that are underlined are correlations of first order constructs on their second order construct. These correlations should not be taken into account for the discriminant validity.

constructs are first order constructs of the second order construct integration of quality management with suppliers by focal firm.

- The highest correlation between constructs is 0.58 (between TSIs by the focal firm and integration of quality management with the focal firm by the buyer) which does not suggest problems of pair wise multicollinearity. Usually correlations between constructs higher than 0.80 result in multicollinearity problems which would prohibit the use of these constructs in one equation (Hair et al., 1998).

7.6.1 The model for the supplier side of the focal firm

In Figure 7.13 the estimated structural model for the supplier side of the focal firm is shown. The overall fit of the model is good, the Chi-square value divided by the number of degrees of freedom is 2.43, which is satisfying, especially regarding the large sample size and model complexity (Baumgartner and Homburg, 1996). This means that the model explains the covariance matrix well. Also the goodness of fit indices, such as NFI, NNFI and CFI are above the threshold levels of 0.90 and also the AGFI is far above the threshold level of 0.80. Only the GFI is slightly below the threshold level of 0.90. For evaluating the goodness of fit of structural models, the NFI and AGFI are especially important fit indices, because they are sensitive to the number of significant paths in the model (Hair *et al.*, 1998). The RMSEA was also below the upper threshold of 0.08. Taking all these fit indices together there is no need to worry for misinterpretations of individual parameter estimations.

From Figure 7.13 it can be seen that many paths turned out to be highly significant, which implies that the nomological validity of the model has been achieved. All four exogenous variables namely external pressure $(\gamma = 0.23; t = 5.53)$, TSIs $(\gamma = 0.37; t = 7.58)$, information exchange by ICT $(\gamma = 0.18; t = 4.46)$ and the quality strategy $(\gamma = 0.24; t = 4.81)$ are positively (significantly) related to the integration of quality management with suppliers. Also commitment of suppliers is (significantly) related to integration of quality management with suppliers. Furthermore, integration of quality management with suppliers is positively significantly related to the commitment of the suppliers $(\beta = 0.54; t = 8.76)$ *and vice versa* $(0.13; t = 2.36)$, enforcement by the focal firm $(\beta = 0.62; t = 9.53)$, satisfaction of the focal firm about the suppliers $(\beta = 0.18; t = 2.33)$ and revenue growth of the focal firm $(\beta = 0.29; t = 3.36)$ as was being hypothesed.

Commitment of the suppliers is positively related to satisfaction of the focal firm with the suppliers $(\beta = 0.43; t = 6.99)$, but not to revenue growth of the focal firm $(\beta = -0.03; t = -0.43)$. Enforcement by the focal firm was neither significant related to satisfaction of focal firm with suppliers $(\beta = -0.08; t = -1.20)$ nor to revenue growth of the focal firm $(\beta = -0.09; t = -1.32)$. The relationship between satisfaction of the focal firm with the suppliers and revenue growth of the focal firm was also investigated, but this was not significant $(\beta = 0.00; t = -0.08)$.

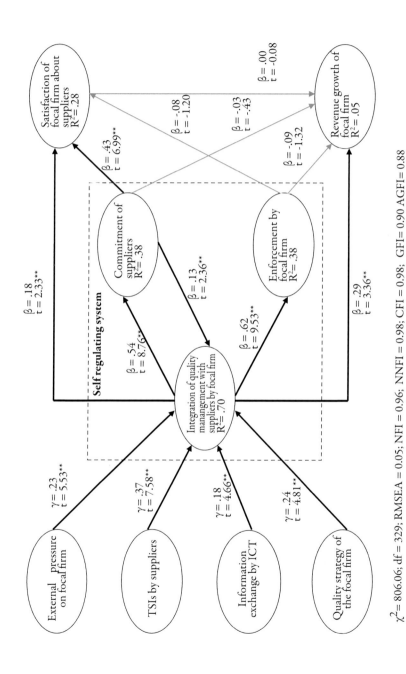

$\chi^2 = 806.06$; df = 329; RMSEA = 0.05; NFI = 0.96; NNFI = 0.98; CFI = 0.98; GFI = 0.90 AGFI= 0.88

** $p \leq 0.01$, * $p \leq 0.05$; N = 540

Figure 7.13. The structural model for the supplier side of the focal firm. Thick lines represent significant relationships.

The explanatory power of the main part of the model was also good; three out of five equations have an R^2 of more than 0.30 which is satisfying in social sciences. The integration of quality management with suppliers by the focal firm in particular is well predicted by the model, with an R^2 of 0.70. Only the revenue growth of the focal firm was not so well explained by the model, most likely because revenue growth is dependent on many other variables, which are not included in the research model.

Lisrel also calculates the indirect effects. Indirect effects are the effects of a construct on a construct via another construct. The sum of the direct and indirect effects is called the total effect[65]. Indirect effects should be used only if there is a theoretical justification for using them, because Lisrel calculates all possible total effects. Within this study the total effect of integration of quality management with the suppliers by the focal firm on the performance constructs, satisfaction of focal firm with suppliers and revenue growth of the focal firm is very interesting from a practical view and therefore reported in Table 7.25. From Table 7.25 it can be concluded that integration of quality management in the supplier model is strongly related to satisfaction of focal firm about suppliers and revenue growth of the focal firm.

A last check for multicollinearity was carried out by conducting a series of five multiple regression analyses in which the integration of quality management with suppliers by the focal firm, commitment of the suppliers, enforcement by the focal firm, satisfaction of the focal firm with suppliers and revenue growth of the focal firm for the supplier model served as dependent variables. The highest Variance Inflation Index (VIF) was 1.61, which was below

Table 7.25. Standardised total effects for the supplier model.

	Satisfaction of focal firm about suppliers	Revenue growth of focal firm
Integration of quality management with supplier by focal firm	0.39** (6.57)	0.24** (4.24)

**$p \leq 0.01$.
T-values of standardised total effect between brackets (two tailed).

[65] For example, the total effect of integration of quality management with suppliers by the focal firm on satisfaction of focal firm about suppliers is calculated in the following way: 0.18 (direct effect of integration of quality management with suppliers by focal firm on satisfaction of focal firm about suppliers) + 0.54* 0.43 (effect of integration of quality management with suppliers by focal firm via commitment of suppliers on satisfaction of focal firm about suppliers) + 0.62*-0.08 (effect of integration of quality management with suppliers via enforcement by focal firm on satisfaction of focal firm about suppliers) + 0.54*0.13*0.43 (due to recursive relationship between integration of quality management and commitment in the model) = 0.39.

the threshold level of 10 and the highest condition index was 19.11 which was lower than the threshold level of 30. During this analysis the effects of a number of control variables were also investigated. Included control variables were size (log number of employees), extra quality systems, presence of a quality manager and presence of a chain leader. The most important reason for using regression was that all control variables were binary variables, except for size. As structural equation modelling is very sensitive to deviations from normality, including these variables into the structural model would lead to a significant drop in the fit of the model[66]. Moreover, it is expected that most control variables have no significant effect on the dependent variables and would only lead to a drop in the fit of the models, because then many relationships are added which do not exist in the data. Because some control variables could also be dependent on the kind of firm, dummy variables for kind of firm were included in the regression analyses[67]. Table 7.26 displays the results[68].

Most control variables have no effect on the dependent variables. Whether or not a firm complies with quality management systems other than the 'standard' quality systems (see also Chapter 2) has a significant impact on the integration of quality management with the suppliers. Specific quality systems might focus on the assurance of certain pressures from the business environment, assuming integration of quality management in the chain as a starting point. Furthermore, the presence of a quality manager is significantly positively related in a strict statistical sense to integration of quality management with suppliers. The quality manager might play an important and supportive role during the integration of quality management systems with suppliers. For the effects of the kind of firm no consistent pattern was observed.

---

[66] In models including one group this could be solved by using the polychloric correlation matrix. However, one should use covariance matrices and not correlation matrices in multigroup analyses (Baumgartner and Homburg, 1996).

[67] One might notice that in the regressions the variance in the dependent variabels is partly explained by the variations due to the sectors. However during the analyses in Lisrel, it was concluded that both the measurement model and the structural model were invariant across the groups. This might seem contradictory, but this is not the case, because in the two analyses different things are tested. In the regression model it is investigated to what extent the independent sector variable contributes to the variations in the dependent variabels. In this case indeed variation in the dependent variables is partly explained by variation of the sector variable. In the Lisrel analyses it is tested if the measurement models and the structural models (β's) are different across the sectors, which is not the case.

[68] The coefficients of the research variables were not included in the table.

Table 7.26. Results of the regression analyses of the control variables on the dependent variables in the supplier model.

	Integration of quality management	Commitment of suppliers	Enforcement by focal firm	Satisfaction of focal firm about suppliers	Revenue growth of focal firm
Supplier model					
Size of focal firm	0.02	-0.05	0.07	-0.09	0.09
Extra management systems	**0.07**	-0.00	0.01	0.03	0.04
Presence of quality manager in focal firm	**0.12**	0.01	0.06	0.01	0.12
Presence of chain leader by focal firm	0.03	0.02	0.04	-0.03	-0.02
*Kind of firm**					
Poultry farmer	**0.16**	0.09	0.04	-0.05	**0.20**
Poultry slaughter/ processor	0.00	**0.19**	0.04	0.03	-0.06
Fruit and vegetable trader/processor	0.07	**0.14**	0.07	-0.08	**-0.11**
Flower and potted plant grower	-0.02	0.09	-0.01	-0.08	0.08
Flower and potted plant trader/ processor	0.01	**0.27**	**0.18**	-0.07	**0.12**
R^2	0.59	0.33	0.27	0.28	0.11

Bold standardised coefficients are significant at $p \leq 0.05$
*Fruit and vegetable growers are used as reference group.

7.6.2 The model for the buyer side of the focal firm

In Figure 7.14 the model for the buyer side of the focal firm is depicted. This model also explains the underlying covariance matrix well, because the value of the Chi-Square divided by the degrees of freedom is 2.83 which is a satisfactory value. The goodness of fit indices NFI, NNFI and CFI are far above the threshold levels of 0.90 and the AGFI is also far above its threshold level of 0.80. Like in the supplier model, the GFI is just slightly below the threshold of 0.90. The RMSEA of the buyer model is 0.06 which is a satisfactory value. The AFGI and

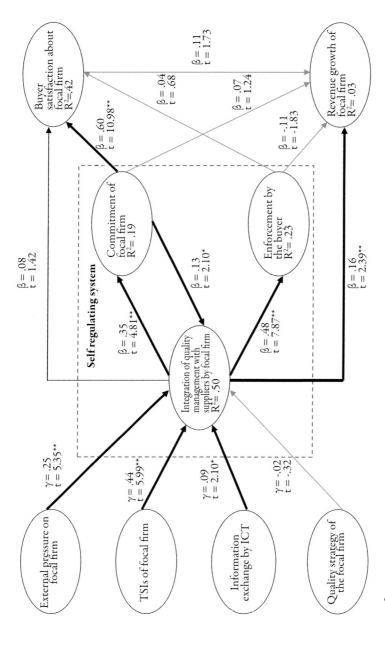

Figure 7.14. The structural model for the buyer side of the focal firm. Thick lines represent significant relationships.

$\chi^2 = 929.45$; df = 329; RMSEA = 0.06; NFI = 0.96 NNFI = 0.97; CFI = 0.97; GFI = 0.89; AGFI = 0.87
$^{**} p \leq 0.01$, $^* p \leq 0.05$; N = 540

the NFI are above their threshold levels; therefore the paths being found in the models can be adequately interpreted.

In the buyer model the external pressure ($\gamma= 0.25; t = 5.35$), TSIs ($\gamma= 0.44; t = 5.99$) and information exchange by ICT ($\gamma= 0.09; t = 2.10$) are strongly related to the integration of quality management with the buyer. For quality strategy no significant effect on the integration of quality management with the buyers was found ($\gamma= -0.02; t = 0.32$). The integration of quality management with the focal firm by the buyer is strongly related to commitment of the focal firm ($\beta= 0.35; t = 4.81$) *and vice versa* ($\beta= 0.13; t = 2.10$) and enforcement by buyers ($\beta= 0.48; t = 7.87$). Commitment of the focal firm is positively related to buyer satisfaction about the focal firm ($\beta= 0.60; t = 10.98$), but not to revenue growth of the focal firm ($\beta= 0.07; t = 1.24$). Enforcement by buyers is not significantly related to both performance indicators ($\beta= 0.04; t = 0.68$ for buyer satisfaction and $\beta=-0.11; t = -1.83$ for revenue growth). Integration of quality management with the focal firm by the buyer does not have a significant direct relationship with buyer satisfaction about the focal firm ($\beta = 0.08; t = 1.42$). However, the direct effect of integration of quality management on revenue growth of the focal firm was also significant ($\beta= 0.16; t = 2.39$). Furthermore, in this model there was a weak significant effect for buyer satisfaction about the focal firm on revenue growth of the focal firm ($\beta= 0.11; t = 1.73$).

The explanatory power for revenue growth of the focal firm in the buyer model is low. The integration of quality management with the buyers is somewhat less explained than in the supplier model. The same holds for commitment. However, buyer satisfaction and enforcement are better explained in the buyer model than in the supplier model.

Table 7.27 displays the standardised total effects of integration of quality management with the focal firm by the buyers on buyer satisfaction with the focal firm and revenue growth of the focal firm, which are both significant.

Five multiple regression analyses (in which the integration of quality management with suppliers with the focal firm, commitment of the suppliers, enforcement by the focal firm,

Table 7.27. Standardised total effects for the buyer model.

	Buyer satisfaction with the focal firm	Revenue growth of focal firm
Integration of quality management with suppliers by focal firm	0.31** (5.21)	0.14** (2.64)

** $p \leq 0.01$.
T-values of standardised effect between brackets (two tailed).

satisfaction of the focal firm about suppliers and revenue growth of the focal firm served as dependent variables) were conducted. A test for multicollinearity was also conducted. Again multicollinearity was not a problem, because the highest VIF was 1.83 and the highest condition index was 27.34, which were both below the threshold levels of respectively 10 and 30. Moreover, during these regression analyses the effects of the control variables were studied. The outcomes of the regression analyses are presented in Table 7.28.

Again including control variables in the analyses did not change the significance and direction of the effects of the research variables, except for a positive effect of integration of quality

Table 7.28. Results of the regression analyses of the control variables on the dependent variables in the buyer model.

	Integration of quality management	Commitment of focal firm	Enforcement by buyer	Satisfaction of buyers with focal firm	Revenue growth of focal firm
Buyer model					
Size of focal firm	-0.10	**0.01**	**0.02**	**-0.07**	**0.11**
Extra management systems	**0.06**	**0.02**	-0.01	**0.03**	**0.05**
Presence of quality manager in focal firm	**0.06**	**0.01**	-0.03	**0.03**	0.12
Presence of chain leader by focal firm	0.08	**0.04**	0.15	**0.04**	**-0.04**
Kind of firm*					
Poultry farmer	**-0.02**	**0.05**	**-0.05**	**-0.01**	0.24
Poultry slaughter/ processor	**-0.05**	**0.06**	0.01	**0.04**	**-0.05**
Fruit and vegetable trader/processor	-0.08	**0.07**	**-0.06**	**-0.03**	-0.08
Flower and potted plant grower	-0.20	**0.07**	**-0.03**	**-0.06**	**0.11**
Flower and potted plant trader/ processor	-0.15	0.20	**-0.04**	**-0.02**	0.13
R^2	0.52	0.17	0.16	0.38	0.11

Bold standardised coefficients are significant at $p \leq 0.05$.
*Fruit and vegetable growers are used as reference group.

management on buyer satisfaction and the effect of buyer satisfaction on revenue growth which was somewhat stronger in the regression analyses[69]. Like in the supplier model, control variables have hardly any effect on each of the dependent variables. Integration of quality management with the buyer by the focal firm has a positive relationship with the presence of a chain leader by the focal firm. Moreover, the revenue growth of the focal firm is positively influenced by the presence of a quality manager in the focal firm. The effect of the kind of firm will be discussed in the next sections. In addition, there is a significant negative relationship found between the size of the focal firm and the integration of quality management with its buyers. An effective integration of quality management requires a lot of efforts per supplier. As the suppliers are larger it may be more difficult for buyers to integrate with them, because of the larger power base of large firms.

7.6.3 In-depth analyses of the external pressures

During the investigation of the impact of external pressures on the integration of quality management an index was used which represents the total pressure. In this section the individual impact on the four drivers from the business environment, *media attention, legislative demands, changing consumer demands* and *societal demands for corporate social responsibility* is determined for both the supplier and buyer model. A regression analyses was carried out in which all the original variables influencing integration of quality management and the control variables were included, however, the pressure of the business environment was replaced by the four individual pressures.

For each drivers an index was created in the same way as this was done for the pressure from the total environment, using the formula presented in Section 5.2.2. These analyses did not show different results with regard to the remaining variables, so only the outcomes for the individual pressures of the business environment on the integration of quality management are shown in Table 7.29. It is remarkable that *legislative demands* have no impact on the integration of quality management, neither with the suppliers nor with the buyers. *Changing consumer demands* only has an impact on the integration of quality management with the buyers as could be expected, because buyers are more close to the end consumers. Negative *media attention* or the inability to comply with *societal demands for CSR* have an impact on integration of quality management, because they can harm the quality reputation of firms which is seen as one of their most important intangible resources (Deephouse, 2000). Media and society exert pressure on firms to conform to public inferences (Greening and Gray, 1994).

---

[69] This is a result of the method used. Differences between Lisrel and SPSS could be the result of taking measurement errors explicitly into account, correlation between errors of indicators, construction of latent variables, extra variables, etc.

Table 7.29. Standardised coefficients from the regression analyses in which the total business environment is spit into the individual pressures.

Pressure	Supplier model	Buyer model
Media attention	**0.09** (2.45)	**0.10** (2.49)
Legislative demands	0.05 (1.32)	0.03 (0.71)
Changing consumer demands	0.04 (1.05)	**0.09** (2.26)
Societal demands for CSR	**0.13** (3.62)	**0.11** (2.60)
R^2	0.58	0.51

Bold standardised coefficients are significant at $p \leq 0.05$.
T-values between brackets (two tailed).

7.6.4 Conclusions regarding the overall supplier and the buyer models

Many paths in the supplier and buyer model show a high similarity with regard to significant relationships. The findings confirm the hypotheses, proposed in Chapter 4 to a large extent. However, there are three important differences between the supplier and buyer model. These will be briefly summarised below:

- The quality strategy of the focal firm is positively related to integration of quality management with the suppliers of the focal firm, but is not significant related to the integration of quality management in the buyer model. This finding supports the vision of this study that integration of quality management in agriculture is initiated by the buyers and not by the suppliers. Suppliers seem not to be able to impose their quality strategy on their buyers.
- Satisfaction of the focal firm with the suppliers is directly dependent on the integration of quality management with the supplier, but satisfaction of the buyer is not significantly dependent on the integration of quality management with the focal firm. A possible explanation can be that respondents have difficulties in providing good estimations of satisfaction of their buyers about their own firm. (Focal firms' respondents answered the questions about satisfaction of their buyers).
- In the buyer model there is a weak significant relationship between buyer satisfaction of the focal firm and revenue growth of the focal firm. However, in the supplier model there is no relationship between satisfaction of the focal firm with the suppliers and revenue growth of the focal firm. A possible explanation can be that the buyers are the ones who pay the firms and buyer satisfaction may lead to repetitive purchases. Satisfaction of the focal firm with the suppliers will not necessary mean that focal firms will realise higher revenue growth.
- When pressures from the business environment are split into its original drivers, it turns out that legislative demands with regard to quality have no impact on the integration of quality management. Media attention and societal demands for corporate social responsibility are important for the integration of quality management, whereas changing consumer demands only has an impact in the buyer model.

7.6.5 Analyses of the sub-groups

In this section the most important results from the regression analyses on the sub-groups of firms are discussed. The sub-groups were compared on the means of the scores of the constructs and the relationships in both the supplier and buyer models. In Appendix 2 the outcomes of the analyses of the sub-groups are described in detail. The most interesting findings of these analyses are presented below.

The scores on the constructs

Although the relationships in the structural model are highly uniform, this does not mean that the levels of scores on the constructs are the same for the different groups. A comparison of the means across groups reveals some interesting results.

1. The scores on the constructs are often significantly lower in the flower and potted plant chain and fewer relationships are significant compared to the poultry meat and the fruit and vegetable chain. Several explanations can be given for this difference:
   - Flower and potted plants are not edible and therefore, the term quality differs compared with the other chains. In food chains, quality mainly deals with food safety, whereas in the flower and potted plant chain, it has to do with environmental and labour issues (see also Table 2.11).
   - For flowers and potted plants fewer regulatory demands exist, except for phytosanitary requirements and the use of pesticides. However, in the two food chains many legislative requirements exist with regard to tracking and tracing, hygiene codes and mandatory participation in monitoring systems of the government (e.g. the Action Plan for the reduction of Salmonella and Campylobacter in the poultry meat chain, see also Section 2.8 and 2.9).
   - Interesting are also the high scores for TSIs of the fruit and vegetable growers in order to comply with the buyer's quality requirements. Recently one of the biggest marketing co-operations enforced an integration of the ICT systems of its growers with its own system, which might explain this high score. This may also explain the high level of information exchange by ICT for primary producers of fruit and vegetables compared to the other chains.
2. The high level of commitment of the suppliers to the quality requirements of poultry farmers compared to other primary producers may be the result of the series of food scandals that have occurred in meat chains in general (see also Box 2.1.).
3. When suppliers (primary producers) and buyers (traders and/or processors) are compared, it seems that suppliers perceive higher TSIs to comply with their buyers' quality requirements than the other way around. Furthermore, buyers are less satisfied with their suppliers as perceived by their suppliers. A more thorough analysis showed that these differences were only present in the fruit and vegetable chain. An explanation for this difference between chains could be that traders and/or processors in the poultry meat and flower and potted plant chain source their products mainly from domestic suppliers, whereas traders and/or

processors of fruit and vegetables source their products from both domestic and foreign suppliers, resulting in less intensive communication. This would be an interesting topic for follow up research.

The relationships in the supplier and buyer models

To a large extent the models of the sub-groups show the same outcomes as the overall models. However, the sub-group analyses revealed some interesting features which are discussed below. It should be noticed that non-significance of relationships in certain groups of firms might be the result of the decrease of statistical power[70].

1. For the primary producers of flowers and potted plants, commitment does not seem to play an important role for integration of quality management, in the buyer model. A potential reason could be that the flower growers mainly deliver their products to the flower auctions in The Netherlands. As a result it is more difficult to directly do business with traders. Therefore, growers may be less aware of the quality requirements of specific traders, which makes it more difficult to be committed to buyers' quality requirements[71].

2. For traders and/or processors in the flower and potted plant chain there is a positive relationship between the integration of quality management and the revenue growth of the focal firms in both models, which is different compared to the other two chains. A likely explanation can be that the level of integration of quality management is significantly lower in this chain compared to the other two chains. Therefore, for traders and/or processors in this chain there are more quick profits to be gained by improving quality management integration. Furthermore, for this group of firms, enforcement and not commitment lead to a higher level of satisfaction about quality by buyers. This is opposite to the general patterns in the data[72]. Because of the lower importance of quality management in the flower and potted plant chain, it is expected that traders and/or processors accept products of growers more easily that do not totally comply with quality requirements. By strong enforcement of quality management requirements, traders might achieve large quality gains.

3. Information exchange by ICT by primary producers in the fruit and vegetable chain is higher compared to the other two chains. However, it has no significant impact on the integration of quality management in the supplier model. A possible explanation is the existence of a kind of threshold level; if information exchange by ICT has reached that level, more frequent use has no further effect on the integration of quality management with suppliers.

---

[70] Statistical power is the probability that one can detect an effect if there really is one. It is influenced by the size of a study. The greater the sample size, the higher the probability that an effect will be detected.

[71] However, in the supplier model there is a signifcant relation between commitment and integration of quality management for primary producers of flowers and plants which is significant at $p \leq 0.10$.

[72] However, suppliers (the growers) do not think that enforcement contributes significantly to the buyer (see customer model of primary producers). This is a likely a difference in perception about the relationship between buyers and suppliers.

4. The significant negative relationship between quality strategy with integration of quality management with buyers for traders and/or processors of fruit and vegetables is remarkable. A possible explanation can be that these traders and/or processors place a lot of emphasis on good quality management in their firms, and in order to preserve their efforts for high quality they expect a strong integration of quality management with their buyers. When these expectations are higher than reality, this may cause a negative perception of the level of the relationship.
5. When perceptions of the suppliers (primary producers) and buyers (traders/processors) about their relationships are compared they agree to a large extent. However, it is especially traders/processors who expect that integration of quality management will help them to achieve a higher revenue growth.

Effects of specific control variables in the sub-group analyses

In the fruit and vegetable chain, young primary producers and the older ones with a successor are more satisfied with the quality performance of their suppliers and achieve a higher revenue growth compared to their main competitors. These firms intend to continue their activities in the future and are doing well, despite the challenges they face in the different agri-food chains. Regarding the variable number of buyers there was a negative significant effect on integration of quality management with the buyers for traders and/or processors in the flower and potted plant chains. A likely explanation is that buyers integrate their quality management with a low number of suppliers in order to reduce their supplier base, a common practice in supply chain management to solve quality problems (Cooper and Ellram, 1993; Kaynak, 2003).

7.7 Concluding remarks

In this chapter the quantitative analysis of responses from 585 firms distributed over six samples were presented. It started with a description of the study samples. Next the analysis of the non-response bias and informant selection was discussed. After that a measurement model including the constructs for both the supplier and the buyer side of the focal firm was developed. Constructs defined in the models were evaluated on metric and factor (co) variance invariance, it turned out that the constructs were highly uniform across the different groups of firms. Also the invariance of the hypotheses across the six groups of firms was empirically supported. Pooling the data together, constructs were tested on content and nomological validity while reflective constructs were in addition tested on convergent validity, discriminant validity and reliability. The outcomes of these tests were highly satisfying. After that, two structural models were developed, one for the supplier side of the focal firm and one for the buyer side. Many of the hypotheses were confirmed and were in line with previous research. Next, several in depth analyses using multiple regression were conducted on sub-groups of firms, which allowed for comparing firms in these groups and for evaluating affects of a number of (sub-group specific) control variables. Again it turned out that the general research model was highly applicable for the different sub-groups and could therefore be regarded as a robust model for modelling quality management and self regulation in agri-food supply chains. The impact of control variables was limited in the different sub-groups. Table 7.30 summarises the outcomes for the hypotheses.

Many authors emphasised the impact of pressures from the business environment on quality management in agri-food supply chains. This study has shown that if firms are facing a strong pressure from the business environment, quality management systems will be more integrated with their suppliers and buyers. Besides pressures from the business environment, also transaction specific investments to comply with the quality requirements of buyers, information exchange by ICT and commitment to quality requirements have a strong positive relationship with the integration of quality management systems. The effect of quality strategy on the integration of quality management was significant for the supplier model, but not for the buyer model, probably because firms with a strong emphasis on quality in their strategy consequently focus their strategy on their suppliers, whereas the firm itself lacks the power to impose its quality strategy on buyers.

Table 7.30. Summary of the hypotheses for both the supplier - and buyer model.

Hypotheses	Expected direction	Relations with suppliers	Relations with buyers
H_1: External pressure → Integration of quality management	+	Supported	Supported
H_2: TSIs → Integration of quality management	+	Supported	Supported
H_3: Information exchange by ICT → Integration of quality management	+	Supported	Supported
H_4: Quality strategy of the focal firm→Integration of quality management	+	Supported	Not supported
H_5: Commitment←→ Integration of quality management	+	Supported	Supported
H_6: Integration of quality management → Enforcement	+	Supported	Supported
H_{7a}: Integration of quality management → Buyer satisfaction	+	Supported	Indirect effect
H_{7b}: Integration of quality management → Revenue growth	+	Supported	Supported
H_{7c}: Commitment → Buyer satisfaction	+	Supported	Supported
H_{7d}: Commitment → Revenue growth	+	Not supported	Not supported
H_{7e}: Enforcement → Buyer satisfaction	0	Supported	Weakly Supported
H_{7f}: Enforcement → Revenue growth	0	Supported	Supported
H_{7g}: Buyer satisfaction → Revenue growth	+	Not supported	Weakly Supported

Within this study, the main elements of the self regulating system are: the integration of the quality management systems combined with the main dimensions of self regulation, commitment and enforcement. The present study reveals that firms in strongly integrated chains show higher levels of compliance behaviour, however, to achieve higher levels of performance firms have to build commitment among their members in the chain and should not use enforcement as a means to condition their supply chain partners.

Chapter 8. In-depth interview results

The in-depth interviews in phase three were used to gain more insight in quality management and to derive 'best practices' in agri-food chains. Furthermore, the interviews were aimed at formulating recommendations to make self regulation work better. Section 8.1 provides the description of the study sample. Section 8.2 discusses the impact of factors on the integration of quality management, while Section 8.3 presents the relationships between integration of quality management and self regulation. In Section 8.4 the impact of integration of quality management and self regulation on performance is described. Section 8.5 provides the concluding remarks.

8.1 Study sample

The interviews were carried out from July to September 2006. Willingness to participate in an interview was asked in a personal phone call in which the objective of the interview was explained. None of the potential experts refused. For each chain involved in this study the objective was to interview at least:
- one primary producer;
- one trader and/or processor;
- one person employed at an interest organisation (e.g. Products Board or trade association);
- one certifier.

The study successfully achieved this objective, however, in the poultry meat sector no certifier was included, because the interviewed poultry farmer had a very broad knowledge of the sector and could easily make comparisons between different quality management systems[73]. Another reason was that two representatives of the Product Boards of Livestock, Meat and Eggs (PVE) were involved, assuring the 'helicopter' view in this chain. In total, fourteen people were interviewed. Their working experience ranged from two years to thirty years, with an average of nineteen years. For these experts quality management was their full time job, assuring that experts were well informed about quality management and self regulation. Firms included in the study were relatively large firms from an agri-food perspective. Those firms often have more resources available to build 'best practice' quality management systems and to deliver products to buyers that have very stringent quality requirements. The outcomes of the in-depth interviews are presented according to the research model in Chapter 3.

---

[73] In the past this poultry farmer was involved in different research activities, has been asked as a speaker on several congresses and attended many meetings in which representatives of the poultry chain were invited.

8.2 Factors influencing integration of quality management

The main factors influencing integration of quality management, as described in the research model were:
- external pressure;
- transaction specific investments (TSIs);
- information exchange by ICT;
- quality strategy.

The outcomes for these factors in the in-depth interviews are discussed below.

8.2.1 External pressure

For most experts in the in-depth interviews it was difficult to give examples of 'best practices' on how to deal with external pressures on individual firms. The business environment was regarded as unpredictable and firms had only limited influence on it. Moreover, most pressures from the business environment were often not targeted at one specific firm but at groups of firms or even a whole sector.

Nine of the fourteen experts mentioned that in many sectors quality management systems exist that deal with specific quality requirements. Compliance with such systems can be regarded as a 'best practice' for dealing with the external pressures. In the fruit and vegetable chain an example is Nature's Choice initiated by the British retailer Tesco. In this system, besides the requirements that are comparable with Eurep-GAP (see Chapter 2), firms must have a plan for managing the business environment, including detailed actions, such as protecting and encouraging wildlife diversity. Pollution control and energy use are also important parts of the scheme, with specific controls on discharges to local watercourses, and energy use reviews by independent third parties. Participating in sector-wide initiatives can be regarded as another 'best practice' for dealing with certain pressures. These initiatives are usually developed by interest organisations or Product Boards and prevent each firm from developing its own approach, resulting in many different quality management systems. For example, three experts from the fruit and vegetable sector mentioned the initiative 'Food Compass' (see also Chapter 2) which helps traders to comply with legislative demands on pesticide residual limits. Two experts mentioned teams in which representatives of the whole sector are involved for dealing with potential crises. In those teams, members are trained to effectively communicate with the media, for example.

Two experts, a trader of fruit and vegetables and a processor of poultry meat were able to give examples about how their firms have directly dealt with external pressures, in their case action groups. In one case the demands of the action groups (Milieudefensie, Natuur en Milieu and Greenpeace) were focused on the compliance with Maximum Residue Limits (MRLs) of pesticides on fruit and vegetables. In the other case demands from the Dutch Animal Welfare

Society were focused on lowering the growing speed of chickens. In both cases the firms co-operated with suppliers and action groups and have launched product-market combinations that take into account the wishes of the action groups. The fruit and vegetable trader had success, the slaughterhouse did not, due to a lack of the consumers' willingness to pay extra for the new product- market combinations.

8.2.2 Transaction specific investments (TSIs)

At least eight experts mentioned TSIs in quality conscious personnel to effectively deal with the specific quality requirements of their buyers. For example, a pepper grower who was delivering to Tesco trained his personnel in the detection, recognition and reporting of harmful insects and plant diseases:

> *Every year a biologist visits our firm to educate our personnel about harmful insects and plant diseases. The early detection leads to decreased use of pesticides resulting in a safer product and less problems with pesticide residuals. This is necessary because our buyer (Tesco) pays a lot of attention to the reduction of pesticide use.*
>
> A pepper grower

Another example was a big fruit and vegetable trader who employs agronomists who visit, and train suppliers abroad to comply with European MRLs. Four experts add that motivation of the employees is important for quality management. Motivated personnel will not leave the firm, so it is not necessary to teach new personnel the quality procedures again and again. Therefore, firms take a lot of effort to find the right personnel, not only as to job requirements, but also to personality and culture.

One expert from the flowers and potted plant sector mentioned the participation in a highly specialised quality system of a grower association as an example of a TSI. Before a grower can fully participate in the quality system a learning trajectory is started. During the learning period the grower pays only 20% of the quality costs. When the grower has reached the desired quality level, an independent certifier visits the firm. If the outcomes of the own audits are comparable to those of the certifier, the grower receives his certificate. Due to the compliance with this system, the grower is able to deliver to British retailers, who place extremely high quality requirements on their suppliers.

8.2.3 Information exchange by ICT

In each chain investigated, large integrated ICT systems exist between buyers and suppliers. Four experts mentioned that especially the large traders and or processors are initiators of these standardised ICT systems. For example, a big fruit and vegetable trader has developed an Internet based system to which suppliers deliver requested quality data to integrate the six different ICT systems of these growers. The outcomes of quality tests and inspections can be (anonymously) published on a web site on which firms can log in and compare their own scores with the scores of others. For some firms benchmarking is an extra stimulus to try to be the best performing firm in their sector.

Five experts emphasised that quality management systems should not be stand alone systems in the chain, but should be integrated with e.g. production management. For example, the quality management system of a slaughterhouse also includes information monitoring, for example, which feed and medicines were used during the growing period of chickens at the farms. Due to this integration of quality management, firms gain more insight in their production processes, what increases and provides more possibilities for improvement, because firms get information more promptly and more frequently.

A group of six experts, including all traders and/or processors, indicated that due to standardisation and integration of quality management easy access to data was obtained in order to answer questions from buyers adequately. As was emphasised by one of the experts:

> *Some time ago a buyer needed data about a specific product within a short time period. The grower was contacted, however, the quality management system was managed by an external person. However, this person was involved in a big sport event at that moment and could not be reached. What would you think about such practices if you were a buyer?*
>
> A pepper grower

8.2.4 Quality strategy

One expert stressed that a firm should have a clear strategy on which position quality management should take relative to the other activities of a firm. If quality management is integrated with the commercial, financial and personnel strategy of the firm, it will more easily be supported by the personnel and it will not be regarded as a bureaucratic burden. If the main interest of a firm is to obtain a quality certificate, but it does not have an appropriate quality strategy, the quality management system will not be successful in the long term. One of the experts made a nice comparison of how the management of a firm should deal with quality management:

> *For these firms, quality management is like a marriage after the wedding day (the day a firm obtains the quality certificate). From that moment the real marriage starts and you have to go for it (applying the quality management system in such a way that it fits within your organisation and with the organisation of the suppliers and the buyers). This will guarantee success in the long run. By doing so, quality management will not be regarded as something that is mandatory and nasty, but as a helpful tool for better firm performance.*
>
> Director of a certification firm

Often awareness of the importance of quality management starts at the strategic level, and goes down through the whole firm or chain. The personnel will find it in their working instructions and the management will have to give good examples to its personnel, like one expert stated:

> *If an employee sees that something goes wrong, resulting in bad quality of products or processes, does he stop to solve the problem, or does he continue his work because he is busy or wants to leave the firm because it is five o'clock? Or just another example; if a buyer demands products that are available, but do not comply with quality requirements, the firm's management should have the discipline to block the delivery.*
>
> Quality manager of flower and potted plant trader

8.3 Self regulating behaviour

The in-depth interviews paid paramount attention to self regulation and its dimensions, *commitment* and *enforcement*. Attention was also paid to the roles of industry organisations, certifying organisations and interest organisations, such as Product Boards and trading associations, in designing self regulated quality management systems.

8.3.1 Commitment

For many traders and/or processors it is a challenge to tie the best performing suppliers on quality to their firm. Three traders and two primary producers emphasise the presence of supplier panels. These panels serve as communication channels between suppliers and buyers and are very useful for increasing commitment as an expert from a big slaughterhouse stated:

> *Every six weeks, we have a meeting with a panel of poultry farmers to discuss topics of quality management ranging from new marketing concepts on quality to the reduction of Salmonella and Campylobacter contamination. We also organise excursions for them to our slaughterhouse. This makes them aware of the consequences of their quality management practices for our quality management. These activities create a lot of commitment for quality management among the poultry farmers.*
>
> Quality manager of a poultry slaughterhouse/processor

Among suppliers in such panels there are vivid discussions of all kind of topics aimed at improving quality performance. For example, a grower that delivers to Tesco organises a monthly meeting with other growers in which novelties, feedback from consumers, information about actions and demanded quantities of products are discussed. Also a newsletter is published every month in which quality management forms are included. Suppliers participating in such panels often come up with improvements and suggestions themselves.

In the 'Table of Eleven' (see Section 3.6), *commitment* consists of the individual dimensions 'knowledge and clarity of regulations', 'the (im) material (dis) advantages of regulations', 'the degree to which regulations are accepted', 'the willingness to comply with non-compliance' and 'the chance of discovering and sanctioning by third parties'. These are discussed in detail below. This discussion reveals the experts' opinions about how self regulation should be improved.

The knowledge and clarity of regulations

Four experts mentioned that most private quality systems in agri-food chains are accredited systems, which means that the regulations and procedures are clearly described and are supervised by an independent Council of Accreditation. According to them, in this way, it is completely clear to firms what the requirements of private quality management systems are. Five experts mentioned that, however, the public regulations from the government were sometimes difficult to understand.

The (im) material (dis) advantages of non-compliance

Firm experts thought they would not gain financially by non-compliance with quality systems, because it could be easily detected by their buyers. Experts like the idea and the emerging practice in which good performing firms on quality are inspected less frequently by the VWA than bad performing firms. If the costs of these controls are charged to firms it introduces a bonus-malus principle for compliance with quality requirements. This approach could increase the motivation of firms to adopt quality regulations, because they know that compliance with the private quality management systems will result in lower inspections. This will remove annoyance, for example because three experts from good performing firms held the opinion that they were inspected too often, although they had an outstanding quality management system.

Three experts mentioned that legislative demands on quality management may result in (logistic) problems in the production processes. An expert from a big poultry processor said that legislative demands on inspections constrained his production process. For example, in The Netherlands a maximum of 9,000 chickens per hour is allowed to be slaughtered, because otherwise controlling agents are not able to control the total flow. This is quite low compared to Belgium and Germany, where 11,000 and 12,000 chickens per hour are slaughtered. It is recommended that government implements quality control systems that are competitive with those of countries abroad[74]. In case of more self regulation such problems may not exist.

---

[74] Recently a pilot study was started in which cameras are placed on the slaughter line to foster control.

The degree to which regulations are accepted

According to six experts self regulation might increase the acceptance of quality regulations. The government often works according to the inspection principle, in which each detail is controlled extensively. This way of working of governmental agencies can be frustrating for firms with good performing quality management systems as an expert complained:

> *These kinds of inspections are disruptive in character, removing initiative and de-motivating. If you trust someone, you do not check everything with a checklist and you do not want to see everything.*

A poultry farmer

Audits make quality requirements more acceptable to firms. If a firm has a certified quality system, the audit is a learning process, in which auditors take the total structure of the firm into account. For example, how processes are organised and how information is communicated within the firm. An auditor is not regarded as a police officer, but as an improver of the firms' processes. Experts would like governmental agencies to work like auditors, because the added value of an auditor is that he helps the entrepreneur to find the balance between the quality requirements and how these requirements should be met within the firm. This difference might partly explain the lack of trust of experts of the VWA in the current quality management systems.

The willingness to comply with regulations

Three experts mentioned that The Netherlands has an important competitive advantage for the introduction of self regulation. In almost all agricultural sectors there is an extensive network of industry organisations, Product Boards and other kinds of associations which represent many firms, whereas many other countries show a lack of organisation of the agricultural sector. These organisations are often active in translating new or changed legislation to their members or introducing initiatives for compliance with quality regulations (e.g. Food Compass), preventing many troubles for firms. This results in a higher willingness to comply with quality regulations.

The chance of discovery and sanctioning by third parties

The VWA expert reported that based on Regulation EU 882/2004[75] it is necessary to be transparent about the outcomes of inspections of controlling agencies. In fact this regulation introduces among others a kind of societal control by the public. The Ministry of Agriculture, Nature and Food Quality and the Ministry of Health, Welfare and Sport have decided that the fruit and vegetable sector will be a pilot sector to work out this EU regulation. The VWA

---

[75] http://eur-lex.europa.eu/LexUriServ/site/nl/consleg/2004/R/02004R0882-20060525-nl.pdf

has launched a web site in which all the results of the residual controls are published with the names of the traders and retailers ('blame-and-shame' approach). If the products fail, it is mentioned on which aspects and whether these aspects are harmful to human health. This approach may increase the efforts of firms to comply with quality management regulations, because they do not want to damage their quality reputation. However, this approach could also have some drawbacks as one of the experts stated:

> *This information is also accessible for Non-Governmental Organisations (NGO) which may use this information in order to put retailers under pressure to come with even more stringent requirements to their suppliers.*
> Representative from an interest organisation in the fruit and vegetable chain

The expert further warns that when the government introduces more frequent inspections for bad performing firms and less frequent inspections for good performing firms, the possibility of finding an offence will become much higher. When these outcomes are published it might seem that the number of non-compliances in The Netherlands is high, especially if these outcomes are used for comparisons of offences across countries.

8.3.2 Enforcement

Regarding enforcement, experts pointed out that almost all big market parties, especially retailers, have summarised their quality requirements for suppliers in certified systems (see Chapter 2). Independent auditors take care of the compliance with such systems and in case of repetitive non-compliance firms will lose their certificate. However, experts warn against using very stringent enforcement principles, because firms that are performing well are hampered by stringent enforcement, which might disillusion and demotivate them. As one of the experts stated:

> *Would it be necessary to develop a very stringent sanctioning system with many controls and inspections for a very small number of firms performing badly and hampering all firms that are performing well? Or would it be better to visit the less good performing firms and to look where the problems occur and to discuss with them how to solve these problems?*
> Quality manager of a poultry slaughterhouse/processor

Moreover, in case of very stringent quality regulations firms might fake compliance with regulations by manipulating measurements. When faking compliance, it might seem that these stringent regulations having some effect, but in practice they do not. If the sanctions are not very severe, it will stimulate timely notification of problems.

The dimensions of *enforcement* of the 'Table of Eleven' are combined in two dimension 'creditability' and 'sanctioning', because they are very close to each other or are difficult to apply individually in agri-food supply chains.

Creditability (chance of control, chance of detection and chance of selection)

The expert from the VWA stated that for the introduction of reliable self regulation it is necessary for certifiers to employ highly qualified independent auditors. Auditors should have the time, knowledge and experience to judge the system on its contents. This enables them to make a judgement on whether all hazards are clearly identified, whether these hazards are really hazards, and whether corrective and preventive actions are needed. According to the expert of the VWA some quality systems deal with the transfer of responsibility to the suppliers and not really with the assurance of quality itself. Another point of attention with regard to the creditability according to this expert was that the government takes care of exact compliance with requirements, whereas in many private quality management systems the certificate is obtained if a firm complies with a certain percentage of the requirements. Therefore, it should be investigated whether partial compliance assures the same level of quality assurance as exact compliance. Three experts from business further stated that the government should realise that fraud is always possible, also with governmental control. For each control organisation, whether it is private or public, it is impossible to check a total firm on its behaviour, as an expert summarised:

> *If people want to do things wrong you can hardly prevent it. If a person wants to use a kind of forbidden pesticide, it is not in the storage of the firm, but in the cabinet at home, or at the neighbours. Inspection agencies will not look in those places.*
>
> Lead auditor of a certification firm

These experts further argued that certifiers know firms and develop relationships with them in which improvement of quality management is very important. As a result they have more insight in the problems and can help to resolve problems. Therefore, they know best whether or not a firm complies well with the quality requirements. Therefore, they may be more effective in preventing fraud because certifiers know firms and develop relationships with them in which improvement of quality management is very important. Experts from business were also aware that introduction of self regulation would not lead to a decrease of the level of quality requirements, but it may even increase the level of the quality regulations, as one expert stated:

> *An important problem with self regulation is that governmental agencies will develop very stringent requirements for self regulation, because they are afraid that something will go wrong if they partly transfer their responsibilities to the market.*
>
> Senior consultant of interest organisation in the flower and potted plant chain

Sanctions (chance and type of sanction)

According to the expert from the VWA the introduction of a Council of Accreditation is a critical success factor for the introduction of self regulation. The commercial relationship between firms and certifiers could hamper certifiers in their sanction possibilities. If a certifier states that a certain firm does not deserve the certificate the firm might go to another certifier. According to an expert from the VWA this could be quite simple for firms, because there is a strong competition between certifiers. However, two experts explicitly mentioned that the common sanction for repeated non-compliance is withdrawal of the certificate or exclusion from delivery, which is much more effective for a firm than a fine as one expert stated:

> Loosing the certificate is often a more rigorous and effective 'shame-and-blame' sanction for a firm than a fine, because a firm is than loosing its market and excluded from the chain, whereas in case of a fine, it can still deliver to the buyer, because the buyer does not know it.
> Lead auditor of a certification firm

Furthermore, retailers themselves are keen to enforce of quality regulations as an expert from the flower and potted plant chain said:

> For the English market it is important to follow exact specifications, although this leads sometimes to very strange situations. For example, the number of fruits on a citrus plant should be in between eight and twelve according to the specifications. In this case the strange situation occurs that extra fruits have to be removed from a plant in order to comply with the specifications.
> Quality manager of a flower and potted plant trader

According to a firm expert the VWA should not worry about the strictness of sanctions in certified quality management systems. However, firm experts warn that the government should take care of firms that operate at the bottom of the market where certificates have no value at all and are only seen as a burden. For some firms the revenues are much higher than the fines: they calculate the fines of the VWA in advance and add them as a budget item in their business administration. One expert of the fruit and vegetable chain guesses that less than one percent of the total trading volume is traded by firms that are not performing well on quality. Big firms in The Netherlands are performing well on quality, but they mainly export their products, whereas bad performing firms often sell their products to the domestic market.

8.4 Performance

Performance was measured by using an operational indicator, *buyer satisfaction* and a financial performance indicator, *revenue growth* of the focal firm. Both are discussed below.

8.4.1 Buyer satisfaction

Eight experts explicitly mentioned that in relationships with strongly integrated quality management systems firms take the initiative to make customised appointments to monitor, to align and to improve the production processes. A flower and potted plant trader gave an example on how this works in practice:

> *For big orders that are placed in advance we visit the growers regularly to make appointments about production according to certain quality specifications. During the growing period we visit them to monitor production, but we also expect growers to notify us in advance in case of problems. Timely notification is important, because it can help us to change our planning schedule.*
>
> Quality manager of a potted plant and flower trader

As already stated, six experts mentioned that firms choose to co-operate with firms that have a similar attitude toward quality management. These firms are often in close contact with each other to achieve common goals (e.g. by means of panels). Within these relationships clear appointments are made about the way these goals should be achieved. An expert stated:

> *Suppliers know to what requirements they have said 'yes' and it prevents expectations (from both sides) that cannot be made true. Discussions at the moment of delivery are annoying, because nothing can be changed at that moment. Most quality systems are nothing else than descriptions how the desired quality should be delivered.*
>
> Quality manager of a flower and potted plant trader

Three experts mentioned that for achieving buyer satisfaction, commitment is much more important than enforcement. If the relationship is good, problems are usually solved in good harmony.

8.4.2 Revenue growth

One expert from a big slaughterhouse was able to give a very specific example on how integration of quality management had a positive impact on the revenue growth and even optimised the total revenues of the whole chain:

> *Due to our extensive registration system we observed that a certain chicken race (A), delivered the biggest quantity of filet per chicken, but took one day more for the farmers to grow in order to obtain the same weight as for other races. Together with a panel of poultry farmers race A was selected and we compensated poultry farmers for the longer growing period of the chickens. As a result the profit of both the slaughterhouse and the chicken farmers grew.*
>
> Quality manager of a poultry slaughterhouse/processor

Four experts had the perception that quality management systems also result in higher costs due to costs for auditing, administration and training as one grower of flowers and potted plants added on the questionnaire:

From 1998 to 2004 I was certified for ISO and FloriMark. The costs and revenues of the systems are totally out of ratio. The buyers buy on basis of trust and not on basis of paperwork.

A flower grower

Quality management systems cost money particularly in the implementation phase, because though firms have the systems they do not yet have the reputation of delivering high quality. Moreover, the exact revenues of good quality management are often difficult to quantify. One expert thought that the relationship between revenue growth and integration of quality management was behaving according to the law of decreasing margins. Thus, if a firm improves from a bad level of quality management, the revenue growth will be great. However, if a firm has already achieved a high quality level, improvements will not contribute much to revenue growth any more. Probably the right answer to the question on whether or not a quality management system results in higher revenues was given by one of the experts:

If you believe in good quality management you see only advantages, but if you do not believe in your quality management system you see only disadvantages.

A pepper grower

8.5 Concluding remarks

This chapter discussed the results of fourteen in-depth interviews that were held with experts in the three chains involved in this study. It provides a number of additional qualitative insights to the quantitative analyses performed in the previous chapter, by focusing on the 'how' and 'why' questions to better explain the relationships found. The questions were open-ended, because then the expert would not be hindered by any framework or bias of the researcher. From these interviews it became clear how firms have implemented self regulating 'best practices' quality management systems in agri-food supply chains and helped to formulate practical recommendations for managers and policy makers. During the interviews it became apparent that it is not always clear what belongs to quality management systems and what not. The boundaries between quality management systems and other systems of the firm are blurring. Information structures that were created for the transfer of information about quality compliance are increasingly used for the transfer of all kinds of product and process related information. One should note that the term 'best practice' quality management system depends on the market a firm is operating in. At the bottom of the market in particular quality regulations are regarded as something firms have to comply with and the perception exists that firms should do as little as possible for the assurance of quality.

Regarding self regulation, the in-depth interviews were useful, because self regulation is quite a broadly defined term in current management and policy research. The discussion of the outcomes of the interviews shed light on the two main dimensions of self regulating behaviour, commitment and enforcement. During the interviews, it turned out that firms would like the implementation of *'control-on-control'* as discussed in Chapter 3 as soon as possible. The main reason was that their quality efforts would be rewarded, because they would get a lower inspection frequency of governmental agencies. In addition, many experts believed that certifiers are able to implement and to supervise the quality management systems better and that certifiers had enough sanction possibilities for firms that tried to cheat the 'rules of the game' in quality management systems. The commercial relationship, the education of the auditors and the necessity of accredited systems were important points of attention to take into account according to the VWA.

The in-depth interviews were complementary to the findings of the questionnaire survey. Whereas the outcomes of the questionnaire were highly uniform, performing a number of in-depth interviews encounters the major drawback of the survey of offering less contextual information. In the next chapter, the conclusions based on the findings of both in-depth interviews and the survey will be drawn, providing recommendations for managers and policy makers to create 'best practices' for self regulated quality management systems in agri-food supply chains.

Chapter 9. Discussion and conclusions

This chapter draws the final conclusions regarding the research questions in Section 9.1. The most important recommendations of this study for managers and policy makers are discussed in Section 9.2. In Section 9.3 the theoretical contributions and in Section 9.4 the methodological implications of the present study are discussed. Section 9.5 sheds light on the limitations of this study and identifies directions for further research. The chapter ends with concluding remarks in Section 9.6.

9.1 Answering of the research questions

The present study investigated the integration of quality management in agri-food supply chains. The most important elements of integration of quality management within agri-food supply chains were defined through various theoretical approaches. Supply Chain Management emphasises the important role of information exchange by ICT for a successful integration of supply chain processes. The importance of external pressures on the way firms should organise the integration of quality management with their buyers and suppliers was underlined by Contingency Theory. Literature indicates that the integration of quality management activities along the supply chain is the best way for firms to deal with these pressures. Transaction Cost Theory indicates the effect of transaction specific investments on the collaboration in supply chains.

This study includes the assumption that chain-wide integration of quality management systems in agri-food supply chains is regarded as the best strategy to deal with today's complex quality demands (Omta *et al.*, 2002). The rationale is that incorrect actions of only one firm may negatively affect the whole supply chain. Closely integrated chains ideally create collaboration in which partners share information, work together to solve problems, jointly plan for the future and make their success interdependent (Krause and Ellram, 1997; Spekman *et al.*, 1998; Shin *et al.*, 2000). By doing so, they effectively achieve a common interest to comply with quality regulations (commitment) and higher transparency in such chains provides firms the means to monitor non-compliance (enforcement). Commitment and enforcement are the two most important dimensions of compliance behaviour in self regulating systems (Balk-Theuws *et al.*, 2004). Finally, it was assumed that self regulated quality management systems would lead to higher performance in agri-food supply chains. This resulted in the following research model (Figure 9.1) that was tested in this study.

The study was carried out in the poultry meat chain, the fruit and vegetable chain and the flower and potted plant chain in The Netherlands and included two successive firms in each chain (primary producers and traders and/or processors). Moreover, the study collected data from both the supplier and buyer side of these firms to ensure the suitable implementation of

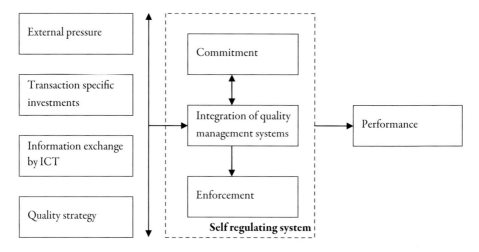

Figure 9.1. Relationships in the research model.

the Supply Chain Management approach. Therefore, the research model has been divided in a supplier and a buyer model.

An interesting finding of this study is that quality management systems were quite comparable across the three chains. Quality management systems in all three supply chains included in this study shared activities such as data exchange of inspection and audit data and evaluation of buyer satisfaction. Standard quality management systems have indeed been developed to cover more than one chain (see Section 2.6). These systems are widely accepted and to a great extent work according to a HACCP approach. In professional networks, such as Product Boards, and producer associations, that span the Dutch agriculture, successful quality management designs may diffuse rapidly, resulting in similar quality management system designs across chains.

Another interesting explanation for the homogeneity of the findings across the three chains is the concept of isomorphism. According to DiMaggio and Powell (1983) isomorphism is a constraining process that forces one unit in a population to resemble other units that face comparable environmental conditions. The institutional perspective of isomorphism places emphasis on the role of social factors including external confirmatory pressures from the business environment, regulatory bodies, buyers and other organisations related to the focal firms in general rather than economical or efficiency factors in driving organisational action (Schuring, 1997; Westphal *et al.*, 1997). The relationship between quality management and isomorphism has been studied in the past in which it was hypothesed (and in many case confirmed) that quality management is fuelled by institutional forces (Westphal *et al.*, 1997; Zbaracki, 1998; Staw and Epstein, 2000; Yeung *et al.*, 2006). Further research should find out whether or not institutional isomorphism is a likely explanation for the high degree of similarity between quality management systems among the three chains[76].

However, scores on individual topics differed significantly across the chains, e.g. the scores were clearly lower in the flower and potted plant chain compared with the poultry meat chain and the fruit and vegetable chain. In food supply chains, quality mainly deals with food safety, whereas in the flower and potted plant chain, it has to do with environmental and labour issues only (see also Table 2.11). As a consequence, for flowers and potted plants fewer regulatory demands with regard to quality exist, except for phytosanitary requirements and the use of pesticides.

9.1.1 Factors influencing the integration of quality management

In order to answer the first research question *Which internal and external factors have an impact on the integration of quality management systems in agri-food supply chains?* a number of factors (external pressure, transaction specific investments, information exchange by ICT and quality strategy) were identified. In Figure 9.2 it is shown to what extent these factors have an impact on the integration of quality management in the supplier and buyer model.

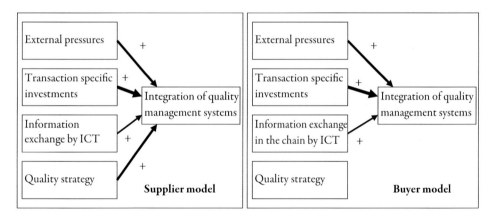

Figure 9.2. Factors influencing integration of quality management with the supplier (left) and buyer (right). The thicker the arrows, the stronger the relationship; no line means no significant relationship.

---

[76] DiMaggio and Powell have identified three forms of isomorphism through which institutional isomorphic change occurs, *coercive, mimic* and *normative* isomorphism:

1. *Coercive* isomorphism occurs through formal and informal pressures exerted on organisations by other organisations which they are dependent on and by expectations in society how organisations should function.
2. *Mimic* isomorphism occurs when firms mimic the actions of successful competitors in the industry.
3. *Normative* isomorphism stems primarily from the professionalisation, which can be defined as the collective struggle of members of an occupation to define the conditions and methods of their work (e.g. Product Board and certifiers).

External pressures

The highly significant relationship between pressures from the business environment and integration of quality management of a firm with its suppliers and buyers supports the notion of Contingency Theory that successful firms adopt more sophisticated governance structures, such as integrated quality management systems, in more uncertain environments. This study confirms the findings of a study of Cap Gemini and Ernst & Young (Grievink *et al.*, 2003) in which a majority of the managers of large European food manufacturers and retailers agrees that assuring quality was a major task for the whole food supply chain. Integration of quality management systems in the supply chain can be regarded as an appropriate strategy to achieve this goal. Many of the pressures included in the present study are not aimed at specific firms, but often influence all firms in the supply chain. However, incorrect actions of only one firm in the supply chain might result in increasing external pressures on all firms in the chain. By integrating quality management in agri-food supply chains managers try to prevent this. The rationale is that in chains in which firms closely work together, the goals of the entire chain become the common objectives of each firm involved. (Lancioni, 2000). From the in-depth interviews it turned out that most firms had no direct contact with the 'sources' where the pressures came from such as the government and action groups. To cope with these pressures, firms adopt specific quality management systems or participate in sector wide quality initiatives (e.g. Food Compass, see Section 2.9.4.). In addition, firms try to participate in marketing channels that fit best with their own quality strategy. For example, firms in the fruit and vegetable chain that are able to deal very well with strict environmental regulations can participate in systems with stringent and far-reaching quality management requirements such as incorporated in 'Nature's Choice' of the British retailer Tesco.

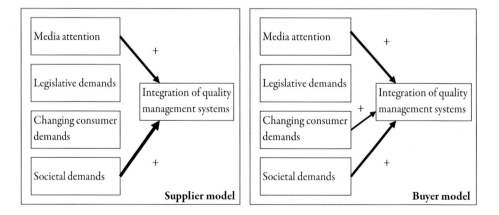

Figure 9.3. Individual pressures from the business environment influencing integration of quality management with the supplier (left) and buyer (right). The thicker the arrows, the stronger the relationship; no line means no significant relationship.

Very interesting facts are highlighted when the construct external pressure is split into the underlying drivers see Figure 9.3. Although some authors, for example, Downey (1996) have recognised *legislative demands* as the most important reason for implementing quality management systems, in this study no significant relationship was found between *legislative demands* and integration of quality management in both the supplier and buyer model. These findings are in line with Gunningham *et al.* (2003) who stated that many firms have complex quality management systems and operate beyond compliance. *Legislative demands* seem to be no longer the primary drivers for the integration of quality management. Two other drivers *media attention* and *societal demands for corporate social responsibility (CSR),* were both found to be significantly related to integration of quality management in both the supplier and the buyer model. Negative *media attention* or the inability to comply with *societal demands for CSR* can harm the quality reputation of firms which is seen as one of their most important intangible resources (Deephouse, 2000). In doing so, the media and society exert pressure on firms to conform to public inference (Greening and Gray, 1994). The fourth driver, *changing consumer demands* was only related to the integration of quality management with downstream buyers. The last finding can be directly related to the growing importance of building consumer oriented chains.

Transaction specific investments (TSIs)

The level of TSIs is strongly related to the integration of quality management in both models. These findings support the logic of Transaction Cost Theory that TSIs are based on and lead to more collaborative governance forms in order to minimise the risk of opportunistic behaviour (Williamson, 1975; Heide and John, 1990; Barney and Hesterly, 1999; Buvik and Halskau, 2001; De Jong and Nooteboom, 2001; David and Han, 2004). When there are many TSIs in the chain, mutual interdependence will occur. The integration of quality management systems creates a shared ground for bilateral multi-strategic control over TSIs. In this study, TSIs have the strongest relationships of all factors with the integration of quality management which supports literature on buyer-supplier relationships. Observing another firm's relation specific investments causes a supply chain member to be more confident in its commitment to the relationship, because the other firm will face economic disadvantages if the relationship ends (Heide and John, 1988; Anderson and Weitz, 1992). Moreover, suppliers with high levels of TSIs and delivering superior benefits to their buyers will be highly valued and buyers will commit themselves to establishing, developing and maintaining relationships with such suppliers (Morgan and Hunt, 1994). From the in-depth interviews on 'best practices' in Chapter 8, human TSIs were mentioned as most frequently applied and most important. Most of the TSIs consisted of training the workforce in order to make them more aware of specific quality demands of buyers.

Information exchange by ICT

Generally speaking, the more the focal firm uses ICT, the more integrated its quality management with its suppliers and/or buyers will be. Quality indicators are measured, stored and transferred in a standardised way. Interestingly, this study shows that the relationship between information exchange by ICT and integration of quality management with suppliers was stronger than with buyers. Leek *et al.* (2003) found opposite results in their study on the use of ICT in the British food sector. According to them, suppliers improve their relationship with buyers by making extensive use of ICT, which might be related to suppliers' desire to provide a better service to their buyers. These findings further support the strong emphasis in Supply Chain Management literature that the enormous development of ICT tools facilitate close co-operation (Cramer, 2004; Matopoulos *et al.*, 2004; Van der Zee, 2004). The integration of information is often the starting point for the successful integration of other processes in supply chains (Hill and Scudder, 2002; Lambert and Cooper, 2000; Swartz, 2000). From the in-depth interviews on 'best practices', it became clear that more standardised and automatic interfaces and flows of data, possibilities that are offered using ICT, allow for more adequate alignment of quality management systems. Moreover, due to the standardisation of data it is possible to carry out detailed analysis on data in order to find the roots of quality problems and to make more 'fact-based' decisions with regard to quality management. Furthermore, it became clear that chain leaders such as retailers or very large buyers are usually initiators for the use of standardised and information exchange by ICT. By doing so, it also allows them to make the quality management systems of various firms in the supply chain more compatible.

Quality strategy of the focal firm

The quality strategy of the focal firm is also positively related to the integration of quality management with suppliers. Successful implementation of quality management systems requires effective change in organisational values/cultures of the firms involved. This is almost impossible without a clear strategy emphasised in the in-depth interviews. Firms that stress quality management in their strategy want to translate their strategy upstream in the chain. For example, firms that have a strong focus on quality may choose quality performance over price when selecting suppliers (Kaynak, 2003; Van der Spiegel, 2004). When the focal firm is a chain leader in particular, it has the (buying) power to impose its quality strategy on its suppliers. However, the quality strategy of the focal firm is not significantly related to the integration of quality management systems with their buyers. Buyers in agri-food supply chains are often much larger then their suppliers and this combined with their buying power, generally makes it very difficult for suppliers to impose their quality strategy on them.

9.1.2 Integrated quality management and self regulating behaviour

The second research question reads *How does integration of quality management affect self regulation?* Regarding this research question, the study has empirically validated that integration of quality management offers good possibilities for successful self regulation of quality assurance. The convergence of quality interests in closely integrated chains leads to a more homogeneous group of firms with a collective interest in good quality management which is seen as a condition for the realisation of self regulation (De Vroom, 1990). Moreover, due the higher level of transparency firms gain more insight in the compliance with each others' quality regulations and have more possibilities for enforcement, which are also regarded as important critical success factors for effective self regulation in agri-food supply chains (Balk-Theuws *et al.*, 2004). The present study formulated two hypotheses assuming that integration of quality management is positively related to commitment (and vice versa) and enforcement.

Commitment

Integration of quality management in the chain is positively related to commitment. Buyers who are quality conscious will primarily choose for suppliers that are also quality conscious. Commitment implies a relationship with the expectation that firms are willing to solve problems together (Morgan and Hunt, 1994; Mehta *et al.*, 2006). During the in-depth interviews many advantages of commitment to quality requirements were mentioned, like focusing on long-term instead of short-term collaboration and a shared vision on quality assurance. Commitment and integration of quality management turned out to be reciprocal. As a relationship evolves over time, mutual acceptance of relational norms forms the basis of future co-operation (Buvik and Halskau, 2001). Open communication on specifications, improvement of processes, transfer of outcomes of specific quality tests and inspections will enhance mutual understanding (Humphreys *et al.*, 2004). This finding is in line with previous research of Kumar *et al.* (1995) who state that commitment will emerge when the interdependence structure is such that the quality interests of the firms in the buyer-supplier relationship converge.

Enforcement

Interestingly, enforcement of quality regulations is also strongly related to integration of quality management in both models. The remark of Grievink *et al.* (2003) that buyers often use integration of quality management as a means to obtain control over their suppliers seems to be confirmed in this study. In particular, large retailers play an active role in quality assurance. These firms can use their power to urge suppliers to participate in quality assurance schemes (Stadifera and Wall, 2003). Therefore, the integration of quality management does not always mean in practice that there has been an alteration of the power balance in agri-food supply chains. Consequently the integration of quality management might increase commitment of firms in the chain, but large buyers retain the ultimate say and may restrict or ration the flow of information in integrated quality management systems.

Regarding the theories dealing with the use of punitive actions, the *relative power theory* seems to provide the largest explanatory power in agri-food supply chains. By integrating their quality management systems, suppliers perceive that their buyers can impose more stringent sanctions in case of non-compliance. Because larger buyers are in general less dependent on their suppliers than the other way around, the possibility that suppliers may apply sanctions it is small. Finally, the questionnaire results reveal that firms that recognise a chain leader on quality in their chain face higher levels of enforcement.

9.1.3 Quality management and performance

In order to answer the third research question: *How do integration and self regulation of quality management systems affect performance in agri-food supply chains?* one main hypothesis was formulated. It stated that integration of quality management would have a positive effect on performance. Furthermore, it was assumed that commitment and not enforcement will show a positive mediating effect between integration of quality management and performance. Performance was measured by using an operational indicator, buyer satisfaction and a financial performance indicator, the revenue growth of the focal firm.

Figure 9.4 shows the different statistical paths between integration of quality management, self regulation and performance variables for the relationship of the focal firm with its suppliers and buyers. The main differences are that in the buyer model, there is not a direct relationship between integration of quality management and buyer satisfaction. Furthermore, in the buyer model a weakly significant relationship between buyer satisfaction and revenue growth exists, which is not present in the supplier model. The relationships are discussed in detail below.

The findings above are largely in line with Morgan and Hunt (1994). These authors have stated that commitment is a key mediating variable for relationship success. The present study showed that commitment was a strong mediator between the integration of quality management and buyer satisfaction. Excessive enforcement is expected to be destructive and destroys successful co-operation in the long-term (Kumar *et al.*, 1995). This turned out to be true in this study, because enforcement is neither related to buyer satisfaction in both models nor to revenue growth in the supplier model and even weakly negatively related to revenue growth in the buyer model.

Integration of quality management with the supplier has a significant positive effect on the satisfaction of the buyer with the quality performance of the focal firm. One of the general characteristics of supply chain integration is that it seeks to fulfil the goal of providing high buyer value with an appropriate use of resources (Cooper *et al.*, 1997; Choi and Eboch, 1998).

Unfortunately, integration of quality management between the focal firm and its buyers does not lead to more satisfaction of the buyers about the quality performance of the focal firm. However, integration of quality management has a positive indirect effect on buyer satisfaction

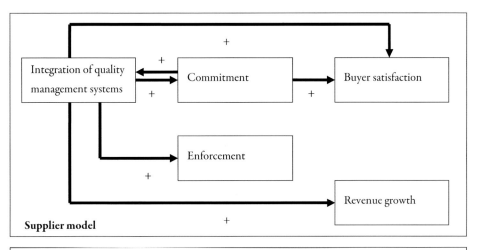

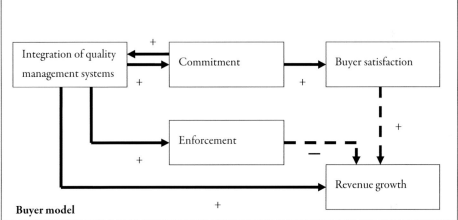

Figure 9.4. Impact of integration of quality management on self regulation and performance of firms in agri-food supply chains (dashed arrow means weak significant relationship).

via commitment. Taking indirect effects in the models into account (see Table 7.25 and 7.27), it turns out that the total effect of integration of quality management on buyer satisfaction in the supplier and buyer models are comparable (see Section 7.6). This means that integration of quality management leads to increased buyer satisfaction, if the focal firm shows strong commitment to the quality requirements of the buyer.

Interestingly, a positive relationship has been found between integration of quality management and revenue growth of the focal firm in both models. Due to the integration of quality management, procedures are developed in which the way of working of the firms is clearly described and monitored. These efforts enlarge the controllability of the production processes of the own firm and facilitate the integration of the processes with those of their suppliers. As a result many quality failures are prevented, resulting in higher efficiencies within the firm and at the supplier or buyer firms. However, the revenue growth is explained only to a small extent by the variables included in the research model (see Figure 7.13 and 7.14). Revenue growth of the firms is dependent on many other factors than those included in this study. Also the temporal aspect of the integration of quality management may play a role here, because quality improvements seldom have an immediate effect on the focal firm's revenue growth (Forza and Filippini, 1998; Singhal and Hendricks, 1999).

Commitment was not significantly related to revenue growth in both models. This means in the supplier model that when the focal firm has suppliers showing a high commitment to quality requirements, it does not necessarily obtain higher revenue growth. For the buyer model this means that if the focal firm shows high level of commitment to quality requirements the revenues do not increase. Regarding the fact that the effect of integration of quality management is larger on buyer satisfaction than on growth of revenues is consistent with Supply Chain Management literature which emphasised that firms seek to fulfil the goals of providing high buyer satisfaction with an appropriate use of resources (Cooper *et al.*, 1997). From a manager's perspective this means that the most important objective is to satisfy buyers, which seems to be more important than the plant's (financial) performance.

Interestingly, both the buyer and the supplier model show that enforcement was not related to buyer satisfaction. Buyers face higher costs to enforce quality requirements by inspections and sanctions. Moreover, enforcement was also not related to revenue growth of the focal firm in the supplier model. Interestingly, in the buyer model, there was a weak but significant *negative* relationship between enforcement and revenue growth. It is likely that firms that fear sanctions for non-compliance do not have a very elaborate quality management system and still have to spend a lot of money in order to comply with the quality requirement of their buyers.

9.2 Managerial and policy implications

In order to answer the fourth and last research question. *What is the best way to create self regulated quality management systems in agri-food supply chains?*, important recommendations are formulated for managers and policy makers.

9.2.1 Implications for managers

The recommendations for managers are focused on establishing 'best practices' in quality management. Managers in agri-food firms might ask themselves questions such as: *How should we start or strengthen the integration of quality management with our buyers and suppliers?, What factors do influence (and how strongly) the integration of quality management with our suppliers and buyers* and *How can we benefit from integration of quality management systems?* The present study has identified a number of important implications from the three phases (conjoint analysis, survey and in-depth interviews) conducted in this study, summarised below.

Jointly dealing with pressures

Managers should realise that in order to deal effectively with pressures from the business environment, firms should integrate their quality management systems with other firms in the chain. The main reason for this is that failure of the quality management system of one firm affects the reputation of all firms in the chain. In strongly integrated supply chains, ideally, the goals of the entire chain become the common objectives of each firm, but also, more information and control actions will become available to the firms in each stage of the supply chain to enforce quality regulations. In this way, integration of quality management increases compliance behaviour. However, managers should realise that they can use their buying power to impose their quality strategy on their suppliers, but it seems to be less possible to impose their quality strategy on their buyers.

Collaboration to improve performance

Managers active in agri-food chains should strive for integration of quality management with their buyers and suppliers, because it is advantageous for firms. It turns out that firms that have higher levels of integration of quality management achieve higher levels of performance in terms of buyer satisfaction and revenue growth. To achieve a high level of performance, and especially buyer satisfaction, it is necessary to find committed parties in the chain that share the firm's objectives with regard to quality management. Commitment can be regarded as the 'glue' that holds together successful buyer supplier relationships. Commitment can be enlarged by:
- Maintaining high quality standards and link up the own firm with exchange partners that have similar visions on quality management.
- Communicating timely quality information and by intensifying the relationships through personal contacts and visits with suppliers and buyers.
- Sharing the benefits from better quality management throughout the supply chain, and stimulating the notification of problems without directly imposing sanctions.

Enforcement of quality requirements should be avoided as much as possible. Strict enforcement does not lead to higher performance in most cases because:
- Strong enforcement of quality regulation has the potential to be destructive and initiates dysfunctional conflict behaviour, especially if sanctions are imposed that are perceived to be unjust or unreasonable. Therefore, stringent enforcement may ruin the necessary (long-term) relationships on quality management in agri-food supply chains.
- If buyers use strong sanctions for non-compliance to quality requirements, suppliers might be faking compliance behaviour and will not notify their buyers in case of quality problems, because they are afraid of stringent sanctions.
- Stringent enforcement by frequent or many controls de-motivates firms that are performing well and results in high and unnecessary monitoring costs for the enforcing firm.

Better use of quality data

Managers should be aware that due to the compliance with quality requirements, they possess a rich source of information about their quality performance over time. At the moment these quality measures are often only used to verify compliance. Analysing this data deeper might reveal the roots of quality problems and indicate ways to solve these problems.

Aligning quality strategy with firm strategy

Managers should develop a clear strategy that positions quality management within the other activities of the firm. If quality management is integrated with the commercial, financial and personnel strategy of the firm, it will not be regarded as a bureaucratic burden and can be better aligned in the firm's processes. Motivation of the personnel could be further improved by developing effective procedures that are short and practical. If the only interest of a firm is to obtain a quality certificate, not supported by an appropriate quality strategy, the quality management system will not be successful in the long run. Therefore, managers should take care that quality management is 'alive' within the firm and should avoid practices that might decrease the perceived importance of good quality management by their employees. Examples are enhancing the quality systems just before auditing or selling products that do not meet the quality requirements in times of product shortage. Such practices may be interpreted by personnel that quality management is just another management fashion, to which they should pay only limited attention.

9.2.2 Implications for policy makers

The study has also derived important implications for policy makers. These recommendations mainly focus on facilitation and improvement of self regulation in agri-food supply chains. Policy makers might be interested in the answers to questions such as: *What hampers the self regulation of quality management in agri-food supply chains?, How can we improve quality management of firms in agri-food supply chains?* and *What is the role of the government with regard to inspections of quality management in the near future?* This study provides the following implications.

Application of 'control-on-control'

Within the concept of *'control-on-control'* the government should retain ultimate responsibility for quality assurance, especially with regard to the mandatory legal European requirements. *'Control-on-control'* is likely to increase the overall level of quality management in agri-food firms. Good performing firms will become even more motivated to improve their quality management, because the governmental inspection frequency of their firms will decrease, lowering the administrative and financial burdens of the inspections. Bad performing firms will be controlled more frequently. Even if the vast majority of firms do the right thing, there is always the chance that irrational, incompetent and stubborn firms will produce serious harm. No less than in other walks of life, there are firms that are simply resistant to new chances. While relatively small and operating at the bottom of the market, these minority of firms cannot be ignored. For such firms, direct inspections offer the most efficient way of ensuring a basic level of quality management. As a result the effectiveness of the governmental inspections will increase, because the government is 'fishing where the fishes are'. The government should stress this higher effectiveness of the *'control-on-control'* approach to other governments in Europe. This is also important in order to avoid unjust comparisons with other EU countries that control all firms. Because bad performing firms are inspected more frequently, it might falsely be suggested that the number of non-compliances in The Netherlands is high. Fair comparisons are extremely important for Dutch agri-food supply chains, because most agri-food chains are highly internationally oriented (see Section 2.2, 2.3, 2.4).

Uniform certification procedures

Policy makers should realise that even the traditional *'command-and-control'* approach of governmental control is not a kind of golden standard for a 100% compliance with quality regulations. Fraud will always be possible. However to minimise the chance of fraud, the procedures should be clearly described and supervised by an independent Council of Accreditation. This prevents the commercial relationship between audited firms and certifiers from hampering certifiers in their evaluation. *'Control-on-control'* might even be more effective in preventing fraud -because certifiers know firms and develop relationships with them in which improvement of quality management is very important. As a result they have more insight in the problems and can help to resolve problems a firm has in complying with regulations better than governmental agencies. Moreover, a common sanction for repeated non-compliance with certified quality management systems is withdrawal of the certificate or exclusion from delivery. This is a severe sanction, because firms that are excluded from their chains no longer have the possibility to deliver to their buyers, which is a more stringent sanction than a fine.

Innovative approaches

The present study showed that not *legislative demands*, but more consumer oriented measures such as increasing *media attention* and *societal demands for corporate social responsibility* have the most important impact on the integration of quality management in agri-food supply chains. Therefore, policy makers should focus on innovative approaches that positively emphasise the efforts of firms to deliver safe and high quality foods, for example:
- Create awards for firms with 'best practice' quality management systems comparable to the corporate social responsibility award of the Ministry of Agriculture, Nature and Food Quality. Winners receive a lot of positive attention in the media.
- Develop a 'score card' including a number of criteria on which the quality management of firms should objectively be assessed. Based on this score card of quality performance a ranking list of firms can be composed and published. It is expected that this ranking

list will start a kind of competition among firms in order to achieve a higher ranking. Of course, this list has to be updated regularly, for example, once a month. This score card could possibly be connected with existing private initiatives such as the Action Plan for Salmonella and Campylobacter and Food Compass (see Section 2.8.4 and 2.9.4).
- Extend the publication of the inspection results on the Internet page of the Dutch Food and Consumer Product Authority to all agri-food sectors, instead of only the fruit and vegetable chain. In order to safeguard their quality reputation firms may intensify the integration of quality management with their suppliers and buyers as a means of complying with quality management regulations (shame-and-blame approach).

Commitment instead of enforcement

Commitment and not enforcement lead to better performance with regard to quality management. This also has consequences for the government. At the moment, the government works according to the inspection principle, which means that many details are checked extensively. Firms perceive this as an enforcement based way of working. In order to change this perception, the government should work according to the auditing principle. The added value of an auditor is that he helps the entrepreneur to find a balance between the quality requirements and the way these requirements should be met within the firm and this is perceived as a commitment based way of working. In this way, the inspection of quality requirements will be perceived more positively by firms.

Retain the advantages of 'control-on-control'

Finally, regarding the possibilities of effective sanctioning in private quality management systems, policy makers should be sure that they do not develop too stringent requirements for self regulating systems if they transfer part of their responsibilities to the market, because they are afraid that something will go wrong. Furthermore, recent research has shown that self regulation in other sectors such as health, higher education and environmental management did not lead to lower administrative burdens for firms (Dorbeck-Jung *et al.*, 2005). Regulations of the government were replaced by all kinds of regulations of private organisations. Because these organisations formulate and implement their regulations from a specific interest, there is less room for own initiatives of firms. Firms face many costs in order to show that they comply with all these new regulations. If this happens, the expected advantages of self regulation will be gone for firms. In order to prevent this, an independent organisation, comparable with a Council of Accreditation or Product Board could judge whether or not these (administrative) regulations are justified.

9.3 Theoretical contribution

The present study has contributed to the development of Supply Chain Quality Management (SCQM) that is regarded by Robinson and Malhotra (2005) as a new stage in the evolution of quality management (see Section 3.2). Up to now quality management and supply chain management have been investigated extensively, but few studies examined these topics jointly. This study has operationalised SCQM and offers a statistically validated and reliable basis for SCQM. This is important, because the new SCQM paradigm needed a reliable conceptual base in order to prevent it from becoming a new management fad (Abrahamson, 1996; Chen and Paulraj, 2004). The usefulness of the SCQM paradigm is further validated, because it was applicable across multiple supply chains and firms within a chain.

Another important contribution is that this study is one among the few quantitative studies which has operationalised the concept of self regulation. By doing so the study was able to contribute to the important debate in buyer supplier relationship literature on the role of commitment and power to establish and to maintain successful long-term relationships (Morgan and Hunt, 1994). The findings of the present study further underline the validity of the key prediction of buyer supplier relationship management described in the influential commitment-trust work of Morgan and Hunt (1994) that commitment is a key mediator variable in achieving successful relationships. The present study showed that closely integrated agri-food supply chains offer promising possibilities for self regulation of quality management. This is an important theoretical contribution, because at the moment most studies have only delivered fragmented and anecdotic information about the relationships between integration of quality management and self regulated quality management systems. Moreover, the study has also shown that self regulated quality management will ultimately result in higher quality and financial performance for the firms involved.

9.4 Methodological implications

The 'mixed methodology' approach used in this study, turned out to be advantageous. The weaknesses of one method have been compensated by the strengths of the others offering a greater potential for consistent theory building. For example, the survey overcomes the limited generalisability of the interviews, while the interviews provide the way of working of firms in their natural setting, things that are hard to include in surveys.

The *first* phase of the study, the Adaptive Conjoint Analysis (ACA) turned out to be very useful as an explorative tool to select the drivers that had the largest impact on quality management in agri-food supply chains. The feature of ACA to evaluate the pressures in relation to each other was very useful, because it mimics the reality of the business environment.

In the *second* phase of the study, a multi-group analysis on both the measurement model and the structural models was performed. This proved to be an appealing methodology for testing measurement equivalence and for investigating invariance hypotheses of substantive interest. It turned out that the findings from the questionnaire were highly generalisable across the six groups of firms included in this study.

The *third* phase of the study, the in-depth interview assisted in gaining feedback on the results and practical insights on how predicted relationships found in phase two actually happen in practice. The in-depth interviews provided a number of 'best practices' about the way quality management and self regulation could be organised and provide recommendations based on these 'best practices'. By doing so, the present study successfully derived a number of useful and important 'take home' messages for managers and policy makers.

Another important methodological implication is that this study has paid substantive attention to implementation of the supply chain perspective throughout the total study. During all the steps described above it collected data from both the supplier and buyer side of the focal firm and includes two successive firms in each chain (primary producers/traders and or processors). Most previous studies are limited to data collection about one firm or only take into account the buyers or the suppliers of the focal firm. Until now such high innovative studies have not been focused on in literature in which data from both suppliers and buyers of multiple firms and multiple chains were gathered.

9.5 Limitations of the study and recommendations for further research

The results and outcomes of this study should be evaluated by taking the following limitations and recommendations into account:

- In this study, quality management is mainly related to process quality and many quality systems (e.g. ISO) are indeed descriptions about the way processes should be carried out. However, for good quality management it is important to take into account product quality as well. Indicators such as costs of waste, time spent on checking, warranty claim costs as percentage of total sales, etc. could be used in further studies to analyse this.
- The questionnaire included questions to measure perceptions from suppliers and buyers, but these questions were answered by the respondents of the focal firm alone. It is questionable to what extent the respondents of the focal firm were able to adequately answer these questions. Further studies could identify the focal firm and its most important supplier and buyer and send the questionnaire to these suppliers and buyers specifically.
- In the questionnaire statements, the buyer was the one who initiated the integration of quality management. In fact it is assumed that the buyer has the power over the supplier to start integration of quality management. Further research about the integration of quality management and self regulation of quality management should pay more attention to the power balances of firms in the chain.
- The present study included the flower and potted plant chain as one chain. However, the flower chain and the potted plant chain show important differences with regard to buyer-supplier relationships. Flowers are mainly sold by the auction clock which implies that most growers do not communicate directly with their buyers. Therefore, growers' (quality) reputation is a very important means of communicating with buyers. In the potted plant chain, deals are made directly between suppliers and buyers. In this channel, the buyer

and the supplier make management decisions about delivery time, quantity and price. The present study has not further investigated this issue.

- In Chapter 2 for each chain a number of quality management systems are described, together with the most important quality legislations in each chain. Many quality management systems stated that they incorporate the legislation which is applied to the firms in which they are operational, for example, Eurep-GAP (see Section 2.6). Because it is difficult to judge whether or not the legislative demands are included in the private systems in an appropriate way, future research should also concentrate on this topic. The main reason is that the legislative requirements are typically of a general nature, so called *open norms*, while the private quality management systems are based on specific requirements, so called *closed norms*. In order to make a reliable judgement whether or not requirements in private quality management cover the legislative quality demands, for each private requirement it is necessary to be investigate whether it is a requirement beyond, equal or below legislative demand.

9.6 Concluding remarks

Today, firms in agri-food supply chains find themselves in turbulent, uncertain and fast changing business environments including a large variety of drivers that exert strong pressures on firms to comply with quality requirements. The integration of quality management systems throughout the supply chain is considered by many authors as the appropriate strategy for effective quality assurance in agri-food supply chains (Omta *et al.*, 2002; Grievink *et al.*, 2003). If quality problems arise and recalls are necessary all parties in an agri-food supply chain will be affected and therefore, all supply chain partners should take their responsibility to assure the quality of food. Moreover, it has been argued that strongly integrated chains are: (1) effective means for establishing self regulating behaviour with regard to quality assurance and (2) due to the higher transparency in such chains, they effectively achieve a common interest to comply with quality regulations (commitment) and have the means to sanction each other in case of non-compliance (enforcement). Finally, many studies have also stressed the positive effect of integration of quality management in buyer-supplier relationships on buyer satisfaction and financial performance (Forza and Filippini, 1998; Flynn and Saladin, 2001; Rungtusanatham *et al.*, 2005).

Until now much attention has been paid to supply chain management, quality management, self regulation and performance but the interlinking between these theories has often been limited and tangential in nature, partly, because of the methodological and practical problems involved. The approach of the present study is that it has provided evidence that confirm the considerations described above by using a 'mixed methodology'. Furthermore, the present study has effectively addressed the Supply Chain Management perspective. During all the steps described above it collected data from both the supplier and buyer side of the focal firm and included two successive firms in three agri-food supply chains (primary producers/traders and/or processors). Taking into account the variation between the chains, the comparison has shown surprisingly consistent results. Consequently, the results may be projected on the whole Dutch agri-food sector.

In order to establish closely integrated chains that take into account the drivers from their business environment, firms should largely invest in human and physical quality assets, make extensive use of ICT in their supply chain relationship and align their quality strategy with suppliers and buyers. Firms have to find committed exchange partners in the chain that share their vision and that have common objectives with regard to quality management. If the integration of quality management is used as a means enforcing quality regulations the gains in performance will be disappointing, because it can have a detrimental effect on buyer-supplier relationships. Regarding the outcomes, managers of firms that intend to integrate their quality management with their buyers and suppliers should first ask themselves a number of questions:

- Are we capable of making optimal use of the integration of quality management systems with our buyers and suppliers?
- Do we have the necessary financial resources for specific investments to comply with public and private quality requirements?
- Is our quality strategy 'alive' in our firm and does it encourage integration of quality management with suppliers and buyers?
- Do we invest sufficiently in ICT systems to make a seamless flow of quality data possible and to gain more insight in our production processes and those of our suppliers and buyers?
- Do our suppliers and buyers share our vision on quality assurance and do they resist attractive short-term alternatives in favour of expected long-term quality benefits?
- How do we find the right balance between building long-term commitment based relationships and the judicious use of enforcement to guarantee a threshold level of quality compliance in our supply chain?
- How can we prove to government that our quality management systems are working effectively in order to receive lower government inspection frequencies on quality management in the future?

As the present study has shown, many firms are at a very acceptable level of integration and self regulation of quality management, but others are weak in some areas which will thus require special attention. Closely integrated chains show a good platform for self regulation on quality. It is further recommended that the government continues the risk-based inspections as proposed in the *'control-on-control'* framework (see Section 2.6). For a successful introduction of self regulation and further improvement of quality management in the chain, the government or governmental agencies should ask the following questions:

- What instruments do we have to determine whether or not private quality management systems overlap with our legislative quality requirements and to what extent?
- Which criteria do we have to separate 'the wheat from the chaff' in order to distinguish between bad and good performing firms?

- What innovative policy instruments do we have to influence the business environment in such a way that it stimulates the integration of quality management in agri-food supply chains?
- How do we adequately communicate the *'control-on-control'* approach to Dutch agri-food firms and more importantly to the EU to assure a level playing field for the Dutch exporting firms?
- Is it possible to change our inspections to a more audit based approach?
- What requirements are needed to select certifiers if we partly transfer our responsibilities to the market?

This study is intended to stimulate the theoretical insights, practical recommendations and methods described in this book to help government and industry to create common goals to come to effective self regulated and integrated quality management in the Dutch agri-food sectors.

References

Abernathy, F.H., J.T. Dunlop, J.H. Hammond and D. Weil, 2000. Retailing and supply chains in the information age. Technology in Society, 22(1): 5-31.

Abrahamson, E., 1996. Management fashion. Academy of Management Review, 21(1): 254-285.

Agarwal, M.K. and P.E. Green, 1991. Adaptive conjoint analysis versus self explicated models: Some empirical results. International Journal of Research in Marketing, 8(2): 141-146.

Ahire, S.L. and P. Dreyfus, 2000. The impact of design management and process management on quality: an empirical investigation. Journal of Operations Management, 18: 549-575.

Ahire, S.L. and D.Y. Golhar, 1996. Quality management in large versus small firms: an empirical investigation. Journal of Small Business Management, 34(2): 1-13.

Ahire, S.L., D.Y. Golhar and M.A. Waller, 1996. Development and Validation of TQM Implementation Constructs. Decision Sciences, 27(1): 23-55.

AID, 2006. 2005 in vogelvlucht. Jaarverslag Algemene Inspectiedienst (in Dutch), Algemene Inspectiedienst, Kerkrade, The Netherlands, 32p.

AID, 2007. Jaarverslag 2006 (in Dutch), Ministry of Agriculture, Nature and Food Quality, General Inspection Agency, Kerkrade, The Netherlands, 42p.

Alleblas, J.T.W. and N.S.P. De Groot, 2000. De Nederlandse glastuinbouw onderweg naar 2020 (in Dutch), LEI, The Hague, The Netherlands, 44p.

Allen, N.J. and J.P. Meyer, 1996. Affective, continuance and normative commitment to the organisation: An examination of construct validity. Journal of Vocational behaviour, 49: 252-276.

Andersen, O. and A. Buvik, 2001. Inter-firm co-ordination: international versus domestic buyer-seller relationships. Omega-International Journal of Management Science, 29(2): 207-219.

Anderson, E. and B. Weitz, 1992. The Use of Pledges to Build and Sustain Commitment in Distribution Channels. Journal of Marketing Research, 29(1): 18-34.

Anderson, J.C. and D.W. Gerbing, 1988. Structural Equation Modeling in Practice: A review and recommended Two-Step Approach. Psychological Bulletin, 103(3): 411-423.

Anderson, J.C. and J.A. Narus, 1990. A model of distributor firm and manufacturer firm working partnerships. Journal of Marketing, 54(42-58).

Anderson, J.C., M. Rungtusanatham and R.G. Schroeder, 1994. A theory of quality management underlying the Deming management method. Academy of Management Review, 19(3): 472-509.

Andrews, R.N.L., 1998. Environmental regulation and business 'self regulation'. Policy Sciences, 31: 177-197.

Antle, J.M., 1999. Benefits and costs of food safety regulation. Food Policy, 24(6): 605-623.

Aramyan, L., C. Ondersteijn, O. Van Kooten and A. Oude Lansink, 2006, Performance indicators in agri-food production chains. In: C. Ondersteijn, J. Wijnands, R. B. M. Huirne and O. Van Kooten (Eds.), Quantifying the agri-food supply chain. Springer, Wageningen, The Netherlands, pp. 49-66.

Armstrong, J.S. and T.S. Overton, 1977. Estimating Nonresponse Bias in Mail Surveys. Journal of Marketing Research, 14(3): 396.

Baarsma, B., F. Felsö, S. Van Geffen, J. Mulder and A. Oostdijk, 2003. Zelf doen? Inventarisatie van zelfreguleringsinstrumenten (in Dutch). 664, SEO, Amsterdam, The Netherlands, 158p.

References

Balk-Theuws, L.W., G.M. Splinter, A.A. Van der Maas, A.G.J.M. Oude Lansink and B.M.J. Van der Meulen, 2004. Zelfregulering van plantgezondheid in de bloemisterij. Verkenning van behoeften en mogelijkheden (in Dutch), LEI, Den Haag, The Netherlands, 76p.

Barney, J.B. and W. Hesterly, 1999, Organizational Economics: Understanding the Relationship between Organizations and Economic Analysis. In: R. C. Steward and C. Hardy (Eds.), Studying Organization. Sage Publications, pp. 109-141.

Baron, R.M. and D.A. Kenny, 1986. The moderator-mediator variable distinction in social psychological research: Conceptual, strategic, and statistical considerations. Journal of Personality and Social Psychology, 51(6): 1173-1182.

Barzel, Y., 2000. The role of contract in quality assurance. Current Agriculture, Food and Resource Issues, 1: 1-10.

Baumgartner, H. and C. Homburg, 1996. Applications of structural equation modeling in marketing and consumer research: A review. International Journal of Research in Marketing, 13(2): 139.

Baumgartner, H. and J.B.E.M. Steenkamp, 1998. Multi-Group Latent Variable Models for Varying Numbers of Items and Factors with Cross National and Longtinudal Applications. Marketing Letters, 9(1): 21-35.

Beamon, B.M., 1998. Supply chain design and analysis: models and methods. International Journal of Production Economics, 19(3/4): 275-292.

Beamon, B.M., 1999. Measuring supply chain performance. International Journal of Operations & Production Management, 19(3): 275-292.

Behr, R.L. and S. Iyengar, 1985. Television news, real world cues, and changes in the public agenda. Public Opinion Quaterly, 49(1): 38-57.

Benschop, A., 1997, Transactiekosten in de Economische Sociologie (in Dutch). University of Amsterdam, Amsterdam, the Netherlands.

Bentler, P.M. and A. Mooijaart, 1989. Choice of structural model via parsimony: A rationale based on precision. Psychological Bulletin, 106(315-317).

Benton, W.C. and M. Maloni, 2005. The influence of power driven buyer/seller relationships on supply chain satisfaction. Journal of Operations Management, 23(1): 1-22.

Bergman, M.E., 2006. The relationships between affective and normative commitment: review and research agenda. Journal of Organisation Behavior, 27: 645-663.

Berkhout, P. and C. Van Bruchem, 2006. Landbouw Economisch bericht 2006 (in Dutch), LEI, The Hague, The Netherlands, 215p.

Bijman, W.J.J., 2002. Essays on Agricultural Co-operatives. Governance structures in fruit and vegetable chains, Erasmus University Rotterdam, Rotterdam, The Netherlands, 185p.

Black, J., 2002, Critical reflections on regulation. Centre for Analysis of Risk and Regulation, London School of Economics, London.

Black, S.A. and L.J. Porter, 1996. Identification of the Critical Factors of TQM. Decision Sciences, 27(1): 1-22.

Bollen, K. and R. Lennox, 1991. Conventional wisdom on measurement: A structrual equation perspective. Psychological Bulletin, 110(2): 305-314.

Bondt, N., S.D.C. Deneux, J. Van der Roest, G.M. Splinter, S.O. Tromp and J.J. De Vlieger, 2005. Nederlandse levensmiddelenketens (in Dutch), LEI, The Hague, The Netherlands, 74p.

Bondt, N., S.D.C. Deneux, I. Van Dijke, O. De Jong, A.J. Smelt, G.M. Splinter, S.O. Tromp and J.J. De Vlieger, 2006. Voedselveiligheid, ketens en toezicht op controle LEI, The Hague, The Netherlands, 94p.

Borch, O.-J., L. Roenning and L.M. Aarseth, 2004, Competitive strategies of small-scale producers in retailer-chain dominated horticulture markets. In: H. J. Bremmers, S. W. F. Omta, J. H. Trienekens and E. F. M. Wubben (Eds.), Sixth international conference on chain and network management in agribusiness and food industry. Wageningen Academic Publishers, Ede, The Netherlands, pp. 481-486.

Boyd, B.K. and J. Fulk, 1996. Executive scanning and perceived uncertainty: A multidimensional model. Journal of Management, 22(1): 1-21.

Bredahl, M. and L. Zaibet, 1995. ISO 9000 in the UK food sector. The Complete European Trade Digest, 3(2): 29-31.

Buvik, A. and K. Gronhaug, 2000. Inter-firm dependence, environmental uncertainty and vertical co-ordination in industrial buyer-seller relationships. Omega-International Journal of Management Science, 28(4): 445-454.

Buvik, A. and O. Halskau, 2001. Relationship duration and buyer influence in just-in-time relationships. European Journal of Purchasing & Supply Management, 7(2): 111.

Buvik, A. and T. Reve, 2001. Asymmetrical deployment of specific assets and contractual safeguarding in industrial purchasing relationships. Journal of Business Research, 51(2): 101-113.

CBS and LEI, 2006. Land - en tuinbouwcijfers 2006, Landbouw economisch instituut and Statistics Netherlands, The Hague, The Netherlands, 264p.

Chen, I.J. and A. Paulraj, 2004. Towards a theory of supply chain management: the constructs and measurements. Journal of Operations Management, 22(2): 119-150.

Choi, T.Y. and K. Eboch, 1998. The TQM Paradox: Relations among TQM practices, plant performance, and customer satisfaction. Journal of Operations Management, 17(1): 59-75.

Churchill, G.A., 1999, Marketing research: Methodological foundations. Dryden Press, Orlando, Florida, United States of America, 1017p.

Clark, T., P.R. Varadarajan and W.M. Pride, 1994. Environmental management: The construct and research propositions. Journal of Business Research, 29: 22-38.

Claro, D.P., 2003. Managing business networks and buyer-supplier relationshops. How information obtained from the business network affects trusts, transactions specific investments, collaboration and performance in the Dutch Potted Plant and Flower Industry, Wageningen University, Wageningen, The Netherlands, 196p.

Conca, J.C., J. Llopis and J.J. Tarí, 2004. Development of a measure to assess quality management in certified firms. European Journal of Operations Research, 156: 683-697.

Cooper, M.C. and L.M. Ellram, 1993. Characteristics of Supply Chain Management and the Implications for the Purchasing and Logistics Strategy. The international Journal of Logistics Management, 4(2): 13-24.

Cooper, M.C., M.D. Lambert and J.D. Pagh, 1997. Supply Chain Management: More than a new name for logistics. The international Journal of Logistics Management, 8(1): 1-13.

Cousins, P.D., 2002. A conceptual model for managing long-term inter-organisational relationships. European Journal of Purchasing & Supply Management, 8(2): 71-82.

References

Cox, A., 1996. Relational competence and strategic procurement management : Towards an entrepreneurial and contractual theory of the firm. European Journal of Purchasing & Supply Management, 2(1): 57-70.

Cramer, J., 2004, Chains and Networks in sustainable business practices. In: T. Camps, P. Diederen, G.J. Hofstede and B. Vos (Eds.), The emerging world of chains and networks. Bridging theory and practice. Reed Business Information, The Hague, The Netherlands, pp. 73-86.

Croom, S., P. Romano and M. Giannakis, 2000. Supply chain management: an analytical framework for critical literature review. European Journal of Purchasing & Supply Management, 6(1): 67-83.

Crosby, P.B., 1979, Quality is free. McGraw-Hill, New York, United States of America.

Cua, K.O., K.E. McKone and R.G. Schroeder, 2001. Relationships between implementation of TQM, JIT, and TPM and manufacturing performance. Journal of Operations Management, 19: 675-694.

Currall, S.C. and A.J. Towler, 2003, Research methods in management and organisational research: toward integration of qualitative and quantitative techniques. In: A. Tashakkori and C. Teddlie (Eds.), Handbook of mixed methods in social and behavioral research. Sage Publications, Thousand Oaks, California, United States of America, pp. 513-526.

David, R.J. and S.-H. Han, 2004. A systematic assessment of the empirical support for transaction cost economics. Strategic Management Journal, 25(1): 39-58.

De Bakker, E., G.B.C. Backus, T. Selnes, M. Meeusen, P. Ingenbleek and C. Van Wagenberg, 2007. Nieuwe rollen, nieuwe kansen (in Dutch), LEI, The Hague, The Netherlands, 104p.

De Haes, E., W. Verbeke, W. Bosmans, R. Januszewska and J. Viane, 2004, Dynamics and interactions in consumer expectations versus producer motivations towards value related aspects in 'superior'quality meat chains. In: H. J. Bremmers, S. W. F. Omta, J. H. Trienekens and E. F. M. Wubben (Eds.), Sixth international conference on chain and network management in agribusiness and food industry. Wageningen Academic Publishers, Ede, The Netherlands, pp. pp. 318-324.

De Jong, G. and B. Nooteboom, 2001. The causality of supply relationships: A comparison between the US, Japan and Europe, Erasmus University of Management (ERIM), Rotterdam, The Netherlands, 34p.

De la Cruz Déniz Déniz, M. and M.K.C. Suárez, 2005. Corporate Social Responsibility and Family Business in Spain. Journal of Business Ethics, 56(1): 27.

De Leeuw, E.D. and J. Segers, 2002, De Vragenlijst (in Dutch). In: J. Segers (Ed.), Methoden voor de maatschappijwetenschappen. Van Gorcum, Assen, The Netherlands, pp. 173-198.

De Vroom, B., 1990, Verenigde fabrikanten. Ondernemersverenigingen van de voedings- en genotmiddelenindustrie tussen achterban en overheid (in Dutch). Wolters-Noordhoff, Groningen, The Netherlands.

Deephouse, D.L., 2000. Media reputation as a strategic resource: An integration of mass communication and resource-based theories. Journal of Management, 26(6): 1091-1112.

Demirbag, M., S.C.L. Koh, E. Tatoglu and S. Zaim, 2006. TQM and market orientation's impact on SMEs' performance. Industrial Management & Data Systems, 106(8): 1206-1228.

Deneux, S.D.C., H.J. Fels-Klerx, S.O. Tromp and J.J. De Vlieger, 2005. Factoren van invloed op voedselveiligheid (in Dutch), LEI, The Hague, The Netherlands, 67p.

Detailhandel, H., 2005. Jaarboek detailhandel 2005/06 (in Dutch), The Hague, the Netherlands.

Diamantopoulos, A. and H.M. Winklhofer, 2001. Index construction with formative indicators: an alternative to scale development. Journal of Marketing Research, 38(May): 269-277.

Dickinson, D.L. and D. Bailey, 2002, A Comparison between US and European Consumer Attitudes and Willingness to Pay for Traceability, Transparency, and Assurance for Pork Products. In: J. H. Trienekens and S. W. F. Omta (Eds.), Fifth International Conference on Chain and Network Management in Agribusiness and the Food Industry. Wageningen Academic Publishers, Noordwijk, The Netherlands, pp. 2229-2237.

Dillman, D.A., 1978, Mail and telephone surveys: the tailored design method. Wiley, New York, United States of America, 325p.

Dillman, D.A., 2000, Mail and internet surveys: The tailored design method. Wiley, New York, United States of America, 480p.

DiMaggio, P.J. and W.W. Powell, 1983. The Iron Cage Revisited: Institutional Isomorphism and Collective Rationality in Organizational Fields. American Sociological Review, 48(2): 147.

Donker, R.A., A.J. Smelt and C.J. Wever, 2000. Kwaliteitszorgsystemen in agroketens voedselveiligheid. Integrale procescontrole en signalering voor dierlijke en plantaardige producten., Expertisecentrum LNV, onderdeel landbouw, Ede, The Netherlands, 52p.

Dorbeck-Jung, B.R., M.J. Oude Vrielink-van Heffen and G.H. Reussing, 2005, Open normen en regeldruk - Een onderzoek naar de kosten en oorzaken van irritaties bij open normen in de kwaliteitszorg (in Dutch). University of Twente, Enschede, The Netherlands.

Dorward, A., 2001. The effects of transaction costs, power and risk on contractual arrangements: a conceptual framework for quantitative analysis. Journal of agricultural economics, 52(2): 59-73.

Doty, D.H., W.H. Glick and P. Huber, 1993. Fit, equifinality, and organizational effectiveness: A test of two configurational theories. The academy of management journal, 36(6): 1196-1250.

Downey, W.D., 1996, The challenge of food and agri products supply chains, 2nd International conference on chain and network management in agri- and food business. Wageningen Academic Publishers, Wageningen, The Netherlands, pp. 3-13.

Easton, G.S. and J. S.L, 1998. The effects of total quality management on corporate performance: an empirical investigation. Journal of Business, 71(2): 253-307.

Eijlander, P., 1993, Zelfregulering en wetgevingsbeleid. In: Ph. Eijlander, P.C. Gilhuis and J.A.F. Peters (Eds.), Overheid of zelfregulering. Alibi voor vrijblijvendheid of prikkel voor actie. W.E.J. Tjeenk Willink, Zwolle, The Netherlands, pp. 129-139.

Ellram, L.M., 1995. Partnering pitfalls and success factors. International Journal of Purchasing and Materials Management, 31(4): 36-44.

Esbjerg, L. and P. Bruun, 2003. Legislation, standarisation, bottlenecks and market trends in relation to safe and high quality food systems and networks in Denmark, MAPP - Centre for research on consumer relations in the food sector. Aarhus School of Business, Aarhus, Denmark, 25p.

Feigenbaum, A.V., 1986, Total Quality Control. McGraw-Hill, New York, the United States of America.

Field, A., 2003, Discovering Statistics using SPSS for windows. Sage Publications, London, United Kingdom, 496p.

Flora Holland, 2007. Samengaan voor de toekomst, Flora Holland, Naaldwijk, The Netherlands, 49p.

Flynn, B.B. and B. Saladin, 2001. Further evidence on the validity of the theoretical models underlying the Baldrige criteria. Journal of Operations Management, 19(6): 617.

Flynn, B.B., R.G. Schroeder and S. Sakakibara, 1995. The impact of quality management practices on performance and competitive advantage. Decision Sciences, 26(5): 659-691.

Forker, L.B., 1997. Factors affecting supplier quality performance. Journal of Operations Management, 15(4): 243-269.

Forker, L.B., S.K. Vickery and C. Droge, 1996. The contribution of quality to business performance. International Journal of Operations & Production Management, 16(8): 44-62.

Fornell, C., 1983. Issues in the application of covariance structure analysis: A comment. Journal of Consumer Research, 9(4): 443-448.

Fornell, C., M.D. Johnson, E.W. Anderson, J. Cha and B.E. Bryant, 1996. The American Customer Satisfaction Index: Nature, Purpose, and Findings. Journal of Marketing, 60(4): 7.

Fornell, C. and D.F. Larcker, 1981. Evaluating Structural Equation Models with Unobservable Variables and Measurement Error. Journal of Marketing Research, 18(1): 39.

Fornell, C., B.-D. Rhee and Y. Yi, 1991. Direct regression, reverse regression, and covariance structure analsysis. Marketing Letters, 20(3): 309-320.

Fortuin, F., 2006. Aligning innovation to business strategy, Wageningen Univeristy, Wageningen, The Netherlands, 187p.

Forza, C. and R. Filippini, 1998. TQM impact on quality conformance and customer satisfaction: A causal model. International Journal of Production Economics, 55(1): 1-20.

Freriks, A.A., 2006, Ketenssystemen in de Agro-Foodsector. Jurdische vormgeving van afspraken binnen de keten en de relatie tot de publieke taken van de overheid (in Dutch). Universtity of Utrecht, The Netherlands, pp. 31.

Frohlich, M.T. and R. Westbrook, 2001. Arcs of integration: an international study of supply chain strategies. Journal of Operations Management, 19(2): 185-200.

Frombrun, C. and M. Shanley, 1990. What's in a name? Reputation building and corporate strategy. The academy of management journal, 33(2): 233-258.

Frugi Venta, 2006. Jaarverslag 2005, Frugi Venta, The Hague, The Netherlands, 49p.

Fuentes-Fuentes, M.M., C.A. Albacete-Saez and F.J. Llorens-Montes, 2004. The impact of environmental characteristics on TQM principles and organizational performance. Omega-International Journal Of Management Science, 32(6): 425-442.

Gambardella, A., 1992. Competitive advantages from in-house scientific research. The US pharmaceutical industry in the 1980s. Research Policy, 21(391-407).

Geelhoed, L.A., 1993, Deregulering, herregulering en zelfregulering. In: Ph. Eijlander, P.C. Gilhuis and J.A.F. Peters (Eds.), Overheid en zelfregulering. Alibi voor vrijblijvendheid of prikkel voor actie, pp. 33-51.

Gerbing, D.W. and J.C. Anderson, 1984. On the Meaning of Within-Factor Correlated Measurement Errors. Journal of Consumer Research, 11: 572-579.

Ghisellla, J.R., J.P. Campbell and S. Zedeck, 1981, Measurement theory for the behavioral science. Freeman, San Francisco, United States of America, 494p.

Ghoshal, S. and P. Moran, 1996. Bad for practice: A critique of the transaction cost theory. Academy of Management Review, 21(1): 13-47.

Ginsberg, A. and N. Venkatraman, 1985. Contigency perspectives of organizational strategy: A critical review of the empirical research. The academy of management review, 10(3): 421-434.

Giraud-Héraud, E., L. Rouached and L.-G. Soler, 2002, Differentiation strategies and product quality in producer-retailer relationships. In: J. H. Trienekens and S. W. F. Omta (Eds.), Proceedings of the fifth international conference on chain and network management in agribusiness and the food industry. Wageningen Academic Publishers, Noordwijk, The Netherlands, pp. 650-663.

Grabosky, P. and N. Gunningham, 1998, The agricultural industry. In: N. Gunningham and P. Grabosky (Eds.), Smart regulation. Designing environmental policy. Oxford University Press, Inc., New York, United States of America, pp. 268-372.

Grandzol, J.R. and M. Gershon, 1997. Which TQM practices really matter: an empirical investigation. Quality Management Journal, 4(4): 43-59.

Green, K.W., R. McGaughey and M. Gasey, 2006. Market orientation and organisational performance. Supply Chain Management, 11(5): 407-414.

Green, P.E., A.M. Krieger and M.K. Agarwal, 1991. Adaptive conjoint analysis: Some caveats and suggestions. Journal of Marketing Research, 28: 215-222.

Greening, D.W. and B. Gray, 1994. Testing a model of organizational response to social and political issues. Academy of Management Journal, 37(3): 467-498.

Grievink, J.-W., L. Josten and C. Valk, 2003, State of the Art in Food. The changing face of the worldwide food industry. Elsevier Business Information, Doetinchem, The Netherlands, 663p.

Grover, V. and M.K. Malhotra, 2003. Transaction cost framework in operations and supply chain management research: theory and measurement. Journal of Operations Management, 21(4): 457-473.

Gunasekaran, A., C. Patel and E. Tirtiroglu, 2001. Performance measures and metrics in a supply chain environment. International Journal Of Operations & Production Management, 21(1-2): 71-87.

Gunasekaren, A., D.K. Macbeth and R. Lamming, 2000. Modelling and analysis of supply chain management systems: an editorial overview. Journal of the Operational Research Society, 51(10): 1112-1115.

Gunningham, N., R. Kagan and D. Thornton, 2003, Shades of green, business regulation and environment. Stanford, University Press, Stanford, United States of America, 206p.

Gunningham, N. and D. Sinclair, 1998, Instruments for environmental protection. In: N. Gunningham and P. Grabosky (Eds.), Smart regulation. Designing environmental policy. Oxford University Press, New York, United States of America, pp. 37-92.

Hair, J.F., R.E. Anderson, R.L. Tatham and W.C. Black, 1998, Multivariate Data Analysis. Prentice Hall International Inc, Upper Sadle River, New Jersey, United States of America, 730p.

Handfield, R.B. and J. Nichols, E. L., 2004. Key issues in global supply base management. Industrial Marketing Management, 33(1): 29-35.

Hardman, P.A., M.A.G. Darroch and G.F. Ortmann, 2002. Improving cooperation to make the South Africa fresh apple export value chain more competitive. Journal of Chain and Network Science, 2(1): 61-73.

Hathaway, S., 1999. The principle of equivalence. Food Control, 10(4-5): 261-265.

Havinga, T., 2006. Private regulations of food safety by supermarkets. Law & Policy, 28(4).

References

HBAG, 2006. Kengetallen Bloemkwekerijproducten, Hoofdbedrijfsschap Agrarische Groothandel, Aalsmeer, The Netherlands, 27p.

Heide, J.B. and G. John, 1988. The role of dependence balancing in safeguarding transaction specific assets in conventional channels. Journal of Marketing, 52: 20-35.

Heide, J.B. and G. John, 1990. Alliances in industrial purchasing: the determinants of joint action in buyer-supplier relationships. Journal of Marketing Research, 27: 24-36.

Henson, S. and J. Caswell, 1999. Food safety regulation: an overview of comtempory issues. Food Policy, 24: 589-603.

Henson, S. and N.H. Hooker, 2001. Private sector management of food safety: public regulation and the role of private controls. International Food and Agribusiness Management Review, 4: 7-17.

Henson, S. and R. Loader, 2001. Barriers to Agricultural Exports from Developing Countries. World Development, 29: 85-102.

Hill, C.A. and G.D. Scudder, 2002. The use of electronic data interchange for supply chain coordination in the food industry. Journal of Operations Management, 20(4): 375-387.

Hingley, M.K., 2005. Power to all our friends? Living with imbalance in supplier-retailer relationships. Industrial Marketing Management, 34(8): 848-858.

Hobbs, J.E., 1996. A transaction cost analysis of quality, traceability and animal welfare issues in UK beef retailing. British Food Journal, 98(6): 16-26.

Hobbs, J.E., 2003. Traceability in Meat Supply Chains. Current Agriculture, Food and Resource Issues, 4: 36-49.

Hogarth-Scott, S. and G.P. Daripan, 2003. Are co-operation and trust being confused with power? An analysis of food retailing in Australia and the UK. International Journal of Retail and Distribution Management, 31(5): 256-267.

Holleran, E., B.M. E. and L. Zaibet, 1999. Private incentives for adopting food safety and quality assurance. Food Policy, 24: 669-683.

Holling, H., T. Melles and W. Reiners, 1998, How many attributes should be used in a paired comparison task? An empirical examination using a new validation approach, Münster, Germany, pp. 1-16.

Hooker, N.H., 1999. Food safety regulations and trade in food products. Food Policy, 24: 653-668.

Horst, H.S., R.B.M. Huirne and A.A. Dijkhuizen, 1996. Eliciting the relative importance of risk factors concerning contagious animal diseases using conjoint analysis: A preliminary survey report. Preventive Veterinary Medicine, 27(3-4): 183-195.

Huber, J.C., D.R. Wittink, J.A. Fiedler and R.L. Miller, 1991, An Empirical Comparison of ACA and Full Profile Judgments, Sawtooth Software Research Paper Series, Sequim, WA, United States of America, pp. 15.

Humphrey, J. and H. Schmitz, 2002. How does insertion in global value chains affect upgrading in industrial clusters. Regional Studies, 36(9): 1017-1027.

Humphreys, P.K., W.L. Li and L.Y. Chan, 2004. The impact of supplier development on buyer-supplier performance. Omega, 32(2): 131-143.

Jahn, G., M. Schramm and A. Spiller, 2004, The trade-off between generality and effectiveniss in certification systems: A conceptual framework. In: H. J. Bremmers, S. W. F. Omta, J. H. Trienekens and E. F. M. Wubben (Eds.), Dynamics in Chains and Networks. Proceedings of the sixth international conference on chain and network management in agribusiness and food industry. Wageningen Academic Publishers, Ede, The Netherlands, pp. 335-343.

Jansen, J.J.P., F.A.J. Van den Bosch and H.W. Volberda, 2006. Exploratory innovation, exploitative innovation, and performance: Effects of organizational antecedents and environmental moderators. Management Science, 52(11): 1661-1674.

Jick, T.D., 1979. Mixing qualitative and quantitative methods: triangulation in action. Administrative Science Quarterly, 24(602-611).

Johnson, G. and K. Scholes, 1999, Exploring Corporate Strategy. Prentice Hall Europe, Hertfordshire, United Kingdom, 560p.

Johnson, R.M., 1987, Adaptive Conjoint Analysis, Proceedings of Sawtooth Software Conference. Sawtooth Software, Ketchum, ID, United States of America, pp. 253-265.

Johnson, R.M., 1991. Comment on 'Adaptive conjoint analysis: some caveats and suggestions'. Journal of Marketing Research, 28: 223-225.

Jöreskog, K.G., 1999. How large can a standardized coefficient be?, 3p.

Jöreskog, K.G. and D. Sörbom, 1996, LISREL 8 structural equation modeling with SIMPLIS command language. Lawrence Erlbaum Associates, Hillsdale, United States of America.

Jouve, 1998. Principles of food safety legislation. Food Control, 9(2-3): 75-81.

Jukes, D., 1995. Food law harmonization within Europe - a learning opportunity. Food Control, 6(5): 283-287.

Juran, J.M., 1986. Quality trilogy. . 9(8): 19-24.

Karmarkar, U. and R. Pitbladdo, 1997. Quality, class and competition. Management Science, 43(1): 27-39.

Katsikeas, C.S., S. Samiee and M. Theodosiou, 2006. Strategy fit and performance consequences of international marketing standardization. Strategic Management Journal, 27(9): 867-890.

Kaynak, H., 2003. The relationship between total quality management practices and their effects on firm performance. Journal of Operations Management, 21: 405-435.

Kelloway, E.K., 1998, Using LISREL for Structural Equation Modeling. A Researcher's Guide. Sage Publications, Thousand Oaks, California, United States of America, 147p.

Kemp, R.G.M., 1999. Managing Interdependence for joint venture success. An empirical study of Dutch international joint ventures, State University of Groningen, Groningen, The Netherlands, 216p.

Kemp, R.G.M., M. Mosselman and A. Van Witteloostuijn, 2004. The perception competition index, Ministry of Economical Affairs, The Hague, The Netherlands, 167p.

Ketchen, J., D. J. and L.C. Giunipero, 2004. The intersection of strategic management and supply chain management. Industrial Marketing Management, 33(1): 51-56.

Ketokivi, M.A. and R.G. Schroeder, 2004. Perceptual measures of performance: fact of fiction? Journal of Operations Management, 22: 247-264.

King, A. and M. Lenox, 2000. Industry self-regulation without sanctions: The chemical industry's responsble care program. Academy of Management Journal, 43(4): 698-716.

References

Klassen, R.D. and L.C. Angell, 1998. An international comparison of environmental management in operations: the impact of manufacturing flexibility in the U.S. and Germany. Journal of Operations Management, 16(2-3): 177.

Klein-Woolthuis, R.J.A., 1999. Sleeping with the Enemy, Trust, dependance and contracts in interorganisational relationsships, University of Twente, Enschede, The Netherlands, 217p.

Krause, D.R., 1999. The antecedents of buying firms' efforts to improve suppliers. Journal of Operations Management, 17(2): 205-224.

Krause, D.R. and L.M. Ellram, 1997. Success factors in supplier development. International Journal of Physical Distribution Logistics Management, 27(1): 39-52.

Krause, D.R., R.B. Handfield and T.V. Scannell, 1998. An empirical investigation of supplier development: reactive and strategic processes. Journal of Operations Management, 17(1): 39-58.

Kumar, K. and H.G. Van Dissel, 1995. Sustainable collaboration: managing conflict and cooperation in interorganisational systems. MIS Quaterly, 20(3): 279-300.

Kumar, K. and H.G. Van Dissel, 1996. Sustainable collaboration: managing conflict and cooperation in interorganisational systems. MIS Quaterly, 20(3): 279-300.

Kumar, N., L.K. Scheer and J.-B.E.M. Steenkamp, 1998. Interdependence, Punitive Capability, and the Reciprocation of Punitive Actions in Channel Relationships. Journal of Marketing Research, 35(2): 225-235.

Kumar, N., L.K. Scheer and J.B.E.M. Steenkamp, 1995. The Effects of Perceived Interdependence on Dealer Attitudes. Journal of Marketing Research, 32(3): 348.

Lai, K.-H. and T.C.E. Cheng, 2003. Initiatives and outcomes of quality management implementation across industries. Omega, 31: 141-154.

Lambert, D.M., M.A. Emmelhainz and G. J.T., 1996. Developing and implementing supply chain partnerships. International Journal of Physical Distribution Logistics Management, 7(2): 35-47.

Lambert, M.D. and M.C. Cooper, 2000. Issues in supply chain management. Industrial Marketing Management, 29: 65-83.

Lambert, M.D., M.C. Cooper and J.D. Pagh, 1998. Supply Chain Management: Implementation Issues and Research Opportunities. The international Journal of Logistics Management, 9(2): 1-19.

Lamming, R., T. Johnsen, J.R. Zheng and C. Harland, 2000. An initial classification of supply networks. International Journal of Operations & Production Management, 20(5-6): 675-691.

Lancioni, R., 2000. New development in supply chian management for the millenium. Industrial Management & Data Systems, 29: 1-6.

Laros, F.J.M. and J.B.E.M. Steenkamp, 2004. Importance of fear in the case of genetically modified food. Psychology & Marketing, 21(11): 889-908.

Lattin, J., D. Carroll and P.E. Green, 2003, Analysing multivariate data. Brooks/Cole Thompson Learning, Pacific Grove, the United States of America, 556p.

Lawrence, P.R. and J.W. Lorsch, 1967, Organization and Environment: Managing differentiation and integration. Harvard University, Boston, United States of America, 279p.

Lazzarini, S.G., G.J. Miller and T.R. Zenger, 2004. Order with some law: Complementary versus substitution of formal and informal arrangement. Journal of Law, Economics and Organisation, 20(2): 261-298.

Leek, S., P.W. Turnbull and P. Naude, 2003. How is information technology affecting business relationships? Results from a UK survey. Industrial Marketing Management, 32(2): 119-126.

Leiblein, M.J., 2003. The choice of organizational governance form and performance: Predictions from transaction cost, resource-based and real options theories. Journal of Management, 29(6): 937-961.

Lejeune, M.A. and N. Yakova, 2005. On characterising the 4 C's in supply chain management. Journal of Operations Management, 23(81-100).

Lindgreen, A. and M. Hingley, 2002, Monitoring buyer-seller relationships: The Tesco scorecard. In: J. H. Trienekens and S. W. F. Omta (Eds.), Fifth international conference on chain and network management in agribusiness and food industry. Wageningen Academic Publisher, Noordwijk, The Netherlands, pp. 165-175.

Lines, R. and J.M. Denstadli, 2004. Information overload in conjoint experiments. International Journal of Market Research, 46(3): 297-310.

Lloyd, T., S. McCorriston, C.W. Morgan and A.J. Rayner, 2001. The impact of food scares on price adjustment in the UK beef market. Agricultural Economics, 25(2-3): 347.

LNV, 2004a. Slim Fruit. Groenten- en fruitkeuringen op een nieuwe manier, Ministry of Agriculture, Nature and Food Quality, The Hague, The Netherlands, 45p.

LNV, 2004b. Toezicht op controle (Toezicht op Toezicht) (in Dutch), Ministry of Agriculture, Nature and Food Quality, The Hague, The Netherlands, 51p.

Loader, R. and J.E. Hobbs, 1999. Strategic responses to food safety legislation. Food Policy, 24: 685-706.

Lu, H., 2007. The role of guanxi in buyer-supplier relationships in China, Wageningen University, Wageningen, The Netherlands, 239p.

Luning, P.A., W.J. Marcelis and W.M.F. Jongen, 2002, Food quality management. Wageningen Academic Publishers, Wageningen, The Netherlands, 323p.

Malhotra, N.K., M. Peterson and S. Bardi Kleiser, 1999. Marketing Research: A state-of-the-Art Review and Directions for the Twenty-First Century. Journal of the Academy of Marketing Science, 27(2): 160-183.

Mark-Herbert, C., 2004, Strategies for collaboration in new product development. In: H. J. Bremmers, S. W. F. Omta, J. H. Trienekens and E. F. M. Wubben (Eds.), 6th International conference on chain and network management in agribusiness and food industry. Wageningen Academic Publishers, Ede, The Netherlands, pp. 48-54.

Masten, S.E., 2000, Transaction-Cost Economics and the organisation of agricultural transactions. In: M. R. Baye (Ed.), Advances in Applied Microeconomics - Industrial Organization. Elsevier Science, New York, United States of America, pp. 173-195.

Matopoulos, A., M. Vlachopoulou, D. Folinas and V. Manthou, 2004, Information architecture framework for agri-food networks. In: H. J. Bremmers, S. W. F. Omta, J. H. Trienekens and E. F. M. Wubben (Eds.), 6th International Conference on Chain and Network Management in Agribusiness and Food Industry. Wageningen Academic Publishers, Ede, The Netherlands, pp. 159-165.

References

Mazé, A., 2002, Quality assurance and contract adaptation in the agri-food sector: Some paradoxes in retailer-producer relationships. In: J. H. Trienekens and S. W. F. Omta (Eds.), Paradoxes in Food Chains and Networks. Proceedings of the fifth international conference on chain and network management in agribusiness and the food industry. Wageningen Academic Publishers, Noordwijk, The Netherlands, pp. 640-649.

Mehta, R., T. Larsen, B. Rosenbloom and J. Ganitsky, 2006. The impact of cultural differences in U.S. business-to-business export marketing channel strategic alliances. Industrial Marketing Management, 35(2): 156.

Melnyk, S.A. and R.B. Handfield, 1998. May you live in interesting times...the emergence of theory-driven empirical research. Journal of Operations Management, 16(4): 311-319.

Meredith, J.R., 1998. Building operations management theory through case and field research. Journal of Operations Management, 16(4): 441-454.

Meuwissen, M.P.M., I.A. Van der Lans and R.B.M. Huirne, 2004, A synthesis of consumer behaviour and chain design. In: Dynamics in Chains and Networks (Ed.), Sixth International Conference on Chain and Network Management inh Agribusiness and the Food Industry. Wageningen Academic Publishers, Ede, The Netherlands, pp. 310-317.

Miles, R.E., C.C. Snow, A.D. Meyer and H.J. Coleman, 1978. Organisational strategy, structure and process. The Academy of Management Journal, 3(3): 546-562.

Miller, D., 1981. Toward new contingency approach: the search for organizational gestalts. Journal of Management Studies, 18: 1-26.

Miller, D., 1988. Relating Porter's business strategies to environment and structure: analysis and performance applications. The Academy of Management Journal, 31(2): 280-308.

Miller, D., 1992. Environmental fit versus internal fit. Organization Science, 3(2): 159-178.

Miller, D. and P. Friesen, 1978. Archetypes of strategy formulation. Management Science, 24(9): 921-933.

Miller, D. and P. Friesen, 1980. Archetypes of Organizational Transition. Administrative Science Quarterly, 25(2): 268-299.

Miller, D. and P.H. Friesen, 1983. Strategy making and Environment: The third link. Strategic Management Journal, 4(3f): 221-235.

Ministerie van Justitie, 2004. De tafel van elf. Een beknopte toets voor de handhaafbaarheid van regels (in Dutch), Ministerie van Justitie, The Hague, The Netherlands, 16p.

Ministerie van Justitie, 2006. De 'Tafel van Elf'. Een veelzijdig instrument, Expertisecentrum Rechtspleging en Rechtshandhaving, The Hague, The Netherlands, 36p.

Mohrman, S.A., R.V. Tenkasi, E.E. Lawler and G.G. Ledford, 1995. Total quality management: practice and outcomes in the largest US firms. Employee Relations, 17(3): 26-41.

Morgan, R.M. and S.D. Hunt, 1994. The Commitment-Trust Theory of Relationship Marketing. Journal of Marketing, 58(3): 20.

Morris, C. and C. Young, 2000. 'Seed to shelf', 'teat to table', barley to beer' and 'womb to tomb': discourses of food quality and quality assurance schemes in the UK. Journal of Rural Studies, 16: 103-115.

Morse, J.M., 2003, Principles of mixed methods and multimethod research design. In: A. Tashakkori and C. Teddlie (Eds.), Handbook of mixed methods in social and behavioral research. Sage Publications, Thousand Oaks, California, United States of America, pp. 189-208.

Motarjemi, Y., M. Schothorst and F. Käferstein, 2001. Future challenges in global harmonization of food safety legislation. Food Control, 12: 339-346.

Murphy, G., B. Trailer and R.C. Hill, 1996. Measuring performance in enterpreneurship reaserch. Journal of Business Research, 36: 15-23.

Muyle, R., 1998. An Emperical Comparison of Three Variants of the AHP and Two Variants of Conjoint Analysis. Journal of Behavioral Decision Making, 11: 263-280.

Neely, A., M. Gregory and K. Platts, 1995. Performance measurement systems design: a literature review and research agenda. International Journal of Operations & Production Management, 15(4): 80-116.

Novoselova, T.A., M.P.M. Meuwissen and R.B.M. Huirne, 2004, New technology adoption in food chains: a review with special reference to GMO applications in livestock production. In: H. J. Bremmers, S.W.F. Omta, J.H. Trienekens and E. F. M. Wubben (Eds.), Sixth International Conference on Chain and Network Management in Agribusiness and the Food Industry. Wageningen Academic Publishers, Ede, The Netherlands, pp. 33-39.

Ollinger, M., J. MacDonald and M. Madison, 2000. Poultry plants lowering the costs and increases variety. Food Review, 23(2): 2-7.

Omta, S.W.F., 1995. Critical success factors in biomedical research and pharmaceutical innovation. The joint impact of management control and contingencies on performance and effectiveness in resarch laboratories in medical faculties, health research institutes and innovative pharmaceutical companies, State University, Groningen, Groningen, The Netherlands.

Omta, S.W.F., J.H. Trienekens and G. Beers, 2002, A framework for the knowledge domain of chain and network science. In: J. H. Trienekens and S. W. F. Omta (Eds.), Paradoxes in Food and Chains and Networks. Fifth international conference on chain and network management in agribusiness and food industry. Wageningen Academic Publishers, Noordwijk, The Netherlands, pp. 13-20.

Opara, L.U. and F. Mazaud, 2001. Food traceability from field to plate. Outlook on Agriculture, 30: 239-247.

Orme, B., 1998, Sample Size Issues for Conjoint Analysis Studies, Sawtooth Software Researh Paper Series, Sequim, WA, United States of America, pp. 9.

Orme, B., 2003, Which Conjoint Method Should I Use?, Sawtooth Software Research Papers Series, Sequim, WA, United States of America, pp. 7.

Orriss, G.D. and A.J. Whitehead, 2000. Hazard analysis and critical control point (HACCP) as a part of an overall qulaity assurance system in international trade. Food Control, 11: 345-351.

Pannekoek, L., O. Van Kooten, R.G.M. Kemp and S.W.F. Omta, 2005. Entrepeneurial innovation in chains and networks in Dutch greenhouse horticulture. The Journal on Chain and Network Science, 5(1): 39-50.

Petersen, B., S. Knura-Deszczka, E. Ponsgen-Schmidt and S. Gymnich, 2002. Computerised food safety monitoring in animal production. Livestock Production Science, 76(3): 207-213.

Peterson, R.A., 1994. A Meta-Analysis of Cronbach's Coefficient Alpha. The Journal of Consumer Research, 21(2): 381-391.

References

Poole, N.D., 1998, Contracts and institutions, Wye College, London, United Kingdom, pp. 36.

Powell, T.C., 1992. Organizational Alignment as Competitive Advantage. Strategic Management Journal, 13(2): 119-134.

Powell, T.C., 1995. Total Quality Management as Competitive Advantage: A Review and Empirical Study. Strategic Management Journal, 16(1): 15.

Powell, W.W., 1990. Neither market or hierarchy - Network forms of organization. Research in Organizational Behavior, 12: 295-336.

Price, M. and E. Chen, 1993. Total quality management in a small high tech company. California Management Review: 96-117.

PVE, 2006. Livestock, Meat and Eggs in the Netherlands 2006, PVE, Zoetermeer, The Netherlands, 56p.

Rabobank, 2002a. De kleur van samenwerking. Ontwikkelingen in de sierteelt: van solisme naar partnership in ketens (in Dutch), Rabobank, Utrecht, The Netherlands, 45p.

Rabobank, 2002b. De smaak van samenwerking (in Dutch), Rabobank, Utrecht, The Netherlands, 38p.

Reardon, T., J.-M. Codron, L. Busch, J. Bingen and C. Harris, 2001. Global change in agrifood grades and standards: Agribusiness strategic responses in developing countries. International Food and Agribusiness Management Review, 2(3/4): 421-435.

Reeves, C.A. and D.A. Bednar, 1994. Defining quality: Alternatives and Implications. The academy of management journal, 19(3): 419-445.

Rindfleisch, A. and J.B. Heide, 1997. Transaction Cost Analysis, Past, Present and Future Applications. Journal of Marketing, 61: 30-54.

Robinson, C.J. and M.K. Malhotra, 2005. Defining the concept of supply chain quality management and its relevance to academic and industrial practice. International Journal of Production Economics, 96: 315-337.

Ruimschotel, D., 1994, De eerste stap. Een kosten-effecten analyse van een integrale periodieke rapportage van de mate van naleving van de beleidsinstrumentele wetgeving (in Dutch). In: Ministerie van Justitie (Ed.). Ministerie van Justitie, The Hague, The Netherlands.

Rungtusanatham, M., C. Forza, R. Filippini and J.C. Anderson, 1998. A replication study of a theory of quality management underlying the Deming method: insight from an Italian context. Journal of Operations Management, 17: 77-95.

Rungtusanatham, M., C. Forza, B.R. Koka, F. Salvador and W. Nie, 2005. TQM across multiple countries: Convergence Hypothesis versus National Specificity arguments. Journal of Operations Management, 23(1): 43.

Saaty, T.L., 1977. A scaling method for priorities in hierarchical structures. Journal of Mathematical Psychology, 15: 234-281.

Salin, V., 2000. Information technology and Cattle-beef supply chains. American Journal of Agricultural Economics, 82(5): 1105-1111.

Samson, D. and M. Terziovski, 1999. The relationship between total quality management practices and operational performance. Journal of Operations Management, 17(4): 393-409.

Saraph, J.V., P.G. Benson and R.G. Schroeder, 1989. An Instrument for Measuring the Critical Factors of Quality Management. Decision Sciences, 20: 810-829.

Sawtooth Software Inc, ACA manual, Sequim, WA, the United States of America.

Scheuren, F., 2004, What is a survey. American Statistical Association, 68p.

Scholten, V.E., 2006. The Early Growth of Academic Spin-offs: Factors influencing the Early Growth of Dutch Spin-off in the Life Scince, ICT and Consulting, Wageningen University and Research Centre, Wageningen, The Netherlands, 212p.

Schoonhoven, C.B., 1981. Problems with contingency theory: Testing assumptions hidden within the language of contingency 'theory'. Administrative Science Quarterly, 26(3): 349-377.

Schuring, R.W., 1997. Procesmodellering van dynamiek in organisaties (in Dutch), University of Twente, Enschede, The Netherlands, 258p.

Shephard, C. and H. Günther, 2006. Measuring supply chain performance: current research and future directions. International Journal of Productivity and Performance Measurement, 55(3/4): 242-258.

Shin, H., D.A. Collier and D.D. Wilson, 2000. Supply management orientation and supplier/ buyer performance. Journal of Operations Management, 18: 317-333.

Siegel, S. and N.J. Castellan, 1988, Nonparametric statistics for the behavioural sciences. McGraw-Hill, New York, United States of America, 399p.

Simon, H.A., 1957, Models of Man, social and rational: Mathematical essays on rational human behavior in social settings. John Wiley, New York, United States of America.

Simpson, D., D. Power and D. Samson, 2005. Greening the automotive supply chain: a relationship perspective, University of Melbourne, Melbourne, Australia, 25p.

Singhal, V.R. and K.B. Hendricks, 1999. The financial justification of TQM. Center for quality of management journal, 8(1): 3-16.

Skytte, H. and N.J. Blunch, 2001. Food retailers' buying behaviour: An analysis in 16 European countries. Journal on Chain and Network Science, 1(2): 133-145.

Slack, N., S. Chambers, C. Harland, A. Harrison and R. Johnston, 1998, Operations Management. Pitman Publishing, London, United Kingdom, 862p.

Smith-DeWaal, C., 2003. Safe food from a consumer perspective. Food Control, 14(2): 75-79.

Snijkers, G., 2002. Cognitive Laboratory Methods: On pre-testing computerised questionnaires and data quality, Utrecht University / Statistics Netherlands, Utrecht / Heerlen, Netherlands.

Snijkers, G., 2004. Hoe weet ik wat ik meet? Het pre-testen van vragenlijsten (in Dutch). STAtOR (Vereniging van Statistiek), 5(1): 4-8.

Sohn, S.Y. and T.H. Moon, 2003. Structural equation model for predicting technology commercialization success index (TCSI). Technological Forecasting and Social Change, 70(9): 885.

Sousa, R. and C.A. Voss, 2002. Quality management re-visited: a reflective review and agenda for future research. Journal of Operations Management, 20: 91-109.

Spekman, R.E., J. Kamauff and N. Myhr, 1998. An empirical investigation into supply chain management: a perspective on partnerships. Supply Chain Management, 3(2): 53-67.

Spekman, R.E., J. Kamauff and J. Spear, 1999. Towards more effective sourcing and supplier management. European Journal of Purchasing & Supply Management, 5(2): 103-116.

Stadifera, R.L. and J.A. Wall, 2003. Managing conflict in B2B e-commerce. Business Horizons, 46(2): 3-13.

References

Staw, B.M. and L.D. Epstein, 2000. What bandwagons bring: Effects of popular management techniques on corporate performance, repution, and CEO pay. Administrative Science Quarterly, 45(3): 523-556.

Steenkamp, J.B.E.M. and H. Baumgartner, 1998. Assessing measurement invariance in cross-national consumer research. The Journal of Consumer Research, 25(1): 78-90.

Steenkamp, J.B.E.M. and H.C.M. Van Trijp, 1991. The use of LISREL in validating marketing constructs. International Journal of Research in Marketing, 1991: 283-299.

Steiner, F., 2004, Formation and Early Growth of Business Webs: Modular Product Systems in Network Markets Physica Verlag, Heidelberg, Germany, 185p.

Stock, J.R. and D.M. Lambert, 2001, Strategic logistics management. McGraw Hill, Boston, United States of America.

Strandholm, K., K. Kumar and R. Subramanian, 2004. Examining the interrelationshop among perceived environmental change, strategic response, managerial characteristics, and organisational performance. Journal of Business Research, 57: 58-68.

Swartz, J., 2000. Changing retail trends, new technologies, and the supply chain. Technology in Society, 22(1): 123-132.

Tabachnick, B.G. and L.S. Fidell, 2001, Using Multivariate Statistics. Allyn & Bacon, Needham Heights, United States of America.

Tacken, G.M.L. and P.L.M. Van Horne, 2006. Handelsstromen van pluimveevlees (in Dutch), LEI, The Hague, The Netherlands, 47p.

Taguchi, G., 1986, Introduction to Quality Engineering: Designing quality in prodcts and process. White Plains, New York, United States of America, 191p.

Tan, K.C., 2001. A framework of supply chain management literature. European Journal of Purchasing & Supply Management, 7(1): 39-48.

Tan, K.C., S.B. Lyman and J.D. Wisner, 2002. Supply chain management: A strategic perspective. International Journal of Operations & Production Management, 22(614-631).

Tashakkori, A. and C. Teddlie, 2003a, Handbook of mixed methods in social & behavioral research. Sage Publications, Thousand Oaks, California, United States of America, 768p.

Tashakkori, A. and C. Teddlie, 2003b, The past and future of mixed methods research: From data triangulation to mixed models design. In: A. Tashakkori and C. Teddlie (Eds.), Handbook of mixed methods design in social and behaviorlal research. Sage Publications, Thousand Oaks, California, United States of America, pp. 617-700.

Tate, R., 1998, An introduction to modeling outcomes in behavioural and social sciences. Burgess, Edina, United States of America.

Taylor, E., 2001. HACCP in small companies: benefit of burden? Food Control, 12: 217-222.

Taylor, W.A. and G.H. Wright, 2003. A longitudinal study of TQM implementation: factors influencing success and failure. Omega, 31(2): 97-111.

Teddlie, C. and A. Tashakkori, 2003, Major issues and controversies in the use of mixed methods in social and behavioral sciences. In: A. Tashakkori and C. Teddlie (Eds.), Handbook of mixed methods in social and behavioral research. Sage Publications, Inc., Thousand Oaks, California, United States of America, pp. 3-50.

Terziovski, M., D. Samson and D. Dow, 1997. The business value of quality management systems certification. Evidence from Australia and New Zealand. Journal of Operations Management, 15(1): 1-18.

Tompkin, R.B., 2001. Interactions between government and industry food safety activities. Food Control, 12: 203-207.

Tourangeau, R. and K.A. Rasinski, 1988. Cognitive processes underlying context effects in attitude measurement. Psychological Bulletin, 103(3): 299-314.

Trienekens, J., W.v. Plaggenhoef, S. Boschma and S. Willlems, 2005. Research agenda on safe and high quality international food chains, ACC, Den Bosch, The Netherlands, 35p.

Trienekens, J.H. and A.J.M. Beulens, 2001a, The implications of EU food safety legislation and consumer demands on supply chain information systems, International Food and Agribusiness Management Association 2001 Agribusiness Forum and Symposium. IAMA, Sydney, Australia.

Trienekens, J.H. and A.J.M. Beulens, 2001b. Views on inter-enterprise relationships. Production Planning and Control, 12(5): 466-477.

Trienekens, J.H. and S.W.F. Omta, 2002, Introduction. In: J.H. Trienekens and S.W.F. Omta (Eds.), Fifth International Conference on Chain and Network Management in Agribusiness and the Food Industry. Wageningen Academic Publishers, Noordwijk, The Netherlands, pp. 1.

Trienekens, J.H. and J.G.A.G. Van der Vorst, 2003, Quality, Safety and Traceability in Food Supply Chains, Lecture notes course Supply Chain Management, Wageningen, The Netherlands, pp. 36p.

Trienekens, J.H. and S. Willems, 2002, Multidisciplinary view on sustainable development of cross border agri supply chains. In: E. Van Amerongen, C. Van der Harg, R. Kruse and S. Pegge (Eds.), The challenge of global chains; integrating developing countries into international chinas, a potential risk or an opportunity. Wageningen Academic Publishers, Wageningen, The Netherlands, pp. 63-68.

Tuncer, B., 2001. From Farm to Fork? Means of Assuring Food Quality. An analysis of the European food quality inititiatives, The international institute for industrial environmental economics, Lund, Sweden, 138p.

Unnevehr, L. and T. Roberts, 2002. Editorial. Food Control, 13: 73-76.

Unnevehr, L.J., 2000. Food safety issues and fresh food product exports from LDCs. Agricultural Economics, 23(3): 231-240.

Valeeva, N.I., 2005. Cost-effectiveness of improving food safety in the dairy production chain, Wageningen University, Wageningen, The Netherlands, 166p.

Valeeva, N.I., M.P.M. Meuwissen, R.H.M. Bergevoet, A.G.J.M. Oude Lansink and R.B.M. Huirne, 2005. Improving Food Safety at the Dairy Farm Level: Farmers' and Experts' Perceptions. Review of Agricultural Economics, 27: 574.

Van Amstel-Van Saane, M., P. Driessen and P. Glasbergen, 2006, Eco-labeling and information asymmetry: a comparison of five eco-labels in the Netherlands, Institutional Mechanisms for Industry Self-Regulation, Harvard, University, Boston, United States of America, pp. 24.

Van den Brink, T.W.M. and F. Van der Woerd, 2004. Industry specific sustainability benchmarks: An ECSF pilot bridging corporate social responsibility with social investments. Journal of Business Ethics, 55: 187-203.

Van den Oever, A., 2005, Wat vindt u van de Nederlandse tomaat?, Algemeen Dagblad.

Van der Fels-Klerx, H.J., H.S. Horst and A.A. Dijkhuizen, 2000. Risk factors for bovine respiratory disease in dairy youngstock in The Netherlands: the perception of experts. Livestock Production Science, 66: 35-46.

Van der Haar, J.W., R.G.M. Kemp and S.W.F. Omta, 2001. Creating Value that Cannot Be Copied. Industrial Marketing Management, 30(8): 627-636.

Van der Meulen, B.M.J. and M. Van der Velde, 2004, Food Safety Law in the European Union. An introduction. Wageningen Academic Publishers, Wageningen, The Netherlands, 270p.

Van der Spiegel, M., 2004. Measuring effectiveness of food quality management, Ponsen & Looijen, Wageningen, The Netherlands, 182p.

Van der Vorst, J.G.A.G., 2000. Effective Food Supply Chains, Generating, modelling and evaluating supply chain scenarios, Wageningen University, Wageningen, The Netherlands, 305p.

Van der Vorst, J.G.A.G., 2004, Supply Chain Management: Theory and Practices. In: T. Camps, P. Diederen, G.J. Hofstede and B. Vos (Eds.), The emerging world of chains and networks. Bridging theory and practice. Reed Business Information B.V., 's Gravenhage, The Netherlands, pp. 105-146.

Van der Vorst, J.G.A.G. and A.J.M. Beulens, 2002. Identifying sources of uncertainty to generate supply chain redesign strategies. International Journal of Physical Distribution Logistics Management, 32(6): 409-430.

Van der Vorst, J.G.A.G., S. Van Dongen, S. Nouguier and R. Hilhorst, 2002. E-business initiatives in food supply chains; Definition and typology of electronic business models. International Journal of Logistics: Research and Applications, 5(2): 119-138.

Van der Zee, H., 2004, Chains, networks and the enabling role of IT. In: T. Camps, P. Diederen, G.J. Hofstede and B. Vos (Eds.), The emerging world of chains and networks. Bridging theory and practice. Reed Business Information, The Hague, The Netherlands, pp. 87-102.

Van Horne, P.L.M., E.B. Oosterkamp, R. Hoste, L.F. Puister and G.B.C. Backus, 2006. Herkomstaanduiding van vlees: nationaal of Europees? (in Dutch), LEI, The Hague, The Netherlands.

Van Kleef, E., L.J. Frewer, G.M. Chryssochoidis, J.R. Houghton, S. Korzen-Bohr, T. Krystallis, J. Lassen, U. Pfenning and G. Rowe, 2006. Perceptions of food risk management among key stakeholders: Results from a cross-European study. Appetite, 47(1): 46-63.

Van Leeuwen, M.G.A., 2006. Het Nederlandse agrocomplex 2006 (in Dutch), LEI, The Hague, The Netherlands, 46p.

Van Plaggenhoef, W. and M. Batterink, 2003. Overview of food quality systems and legislation for the fruit, beef and fish chain in the Netherlands, Wageningen University, Wageningen, The Netherlands, 32p.

Van Plaggenhoef, W., M. Batterink and J.H. Trienekens, 2003. International Trade and Food Safety: Overview of legislation and standards, EU concerted action Global Food Network, Wageningen, The Netherlands, 52p.

Van Schaik, G., A.A. Dijkhuizen, R.B.M. Huirne and G. Benedictus, 1998. Adaptive conjoint analysis to determine perceived risk factors on farmers, veterinarians and AI technicans for introduction of BHV1 to dairy farms. Preventive Veterinary Medicine, 37: 101-112.

Venkatraman, N., 1989. The concept of fit in strategic coalignment: a methological perspective. Journal of Management Studies, 14(3): 423-444.

Venkatraman, N. and J.E. Prescott, 1990. Environment-Strategy Coalignment: An empirical test of its performance implications. Strategic Management Journal, 11(1): 1-23.

Verbeke, W. and J. Viane, 2002, Demand-oriented Meat Chain Management: The Emerging Role of Traceability and Information Flows. In: J.H. Trienekens and P.J.P. Zuurbier (Eds.), Fourth International Conference on Chain and Network Management in Agribusiness and the Food Industry. Wageningen Academic Publishers, Noordwijk, The Netherlands, pp. 391-400.

Verbeke, W. and R.W. Ward, 2001. A fresh meat almost ideal demand system incorporating negative TV press and advertising impact. Agricultural Economics, 25(2-3): 359.

Verduijn, T.M., 2004. Dynamism in supply networks. Actor switching in a turbulent business environment, Erasmus University, Rotterdam, The Netherlands, 341p.

Verschuren, P. and H. Doorewaard, 1999, Designing a research project. Lemma B.V., Utrecht, The Netherlands, 215p.

VWA, 2004. Toezichsarrangementen risicogebieden diervoederketen (in Dutch), VWA, The Hague, The Netherlands, 36p.

VWA, 2006. Handhaven met verstand en gevoel. Vernieuwing van het handhavingsbeleid van de VWA op basis van risico's en proportionaliteit (in Dutch), VWA, The Hague, The Netherlands, 23p.

VWA, 2007a. Jaarverslag 2006 (in Dutch), VWA, The Hague, the Netherlands, 68p.

VWA, 2007b. Tussenrapportage 'Eenduidig Toezicht' in het domein 'De Vleesketen' (in Dutch), VWA, the Hague, The Netherlands, 67p.

Wacker, J.G., 1998. A definition of theory: research guidelines for different theory-building research methods in operations management. Journal of Operations Management, 16(4): 361-385.

Werts, C., R.L. Linn and K.G. Joreskög, 1974. Intraclass reliablilty estimates: Testing structural assumptions. Educational and Psychological Measurement, 34(25-33).

Westerman, E., B. Boertjes, A. Oude Lansink and L.W. Balk-Theuws, 2005. Herziening van de status van quarantaine organismen. Zijn Liriomyza trifolii en Liriomyza huidobrensis nog wel quarantainewaardig?, LEI, The Hague, The Netherlands, 76p.

Westphal, J.D., R. Gulati and S.M. Shortell, 1997. Customization or Conformity? An Institutional and Network Perspective on the Content and Consequences of TQM Adoption. Administrative Science Quarterly, 42(2): 366.

Wijnands, J. and H.J. Silvis, 2000. Onderweg, concurrentiepositie Nederlandse Agrosector, LEI, The Hague, The Netherlands, 112p.

Williamson, O.E., 1975, Markets and Hierarchies: Analysis and Antitrust Implication. The Free Press, New York, United States of America.

Williamson, O.E., 1985, The economic institutions of capitalism. The Free Press, New York, United States of America, 450p.

Williamson, O.E. (Ed.), 1989. Transaction Cost Economics. Handbook of industrial organisation, 1. Elsevier, Amsterdam, The Netherlands.

Williamson, O.E., 1991. Comparative economic organization: The analysis of discrete structural alternatives. Administrative Quarterly, 36: 269-296.

References

Williamson, O.E., 1996, The mechanism of governance. Oxford University Press, New York, United States of America.

WUR Projectgroep Veerkrachtige Pluimveevleesproductie, 2004. Verkenningen voor een Veerkrachtige Pluiveevleesproductie (in Dutch), Wageningen University, Wageningen, The Netherlands, 85p.

Yasai-Ardekani, M. and P.C. Nystrom, 1996. Designs for environmental scanning systems: Tests of contigency theory. Management Science, 42(2): 187-204.

Yeung, A.C.L., T.C.E. Cheng and K.H. Lai, 2006. An operational and institutional perspective on total quality management. Production And Operations Management, 15(1): 156-170.

Yin, R.K., 1994, Case study research: design and methods. Sage Publications, London, United Kingdom.

Zaheer, A. and N. Venkatraman, 1995. Relational governance as an interorganzational strategy: An empirical test of the role of trust in economic exchange. Strategic Management Journal, 26(5): 373-392.

Zajac, E.J., M.S. Kraatz and R.K.F. Bresser, 2000. Modeling the dynamics of strategic fit: A normative approach to strategic change. Strategic Management Journal, 21(4): 429-453.

Zajac, E.J. and S.M. Shortell, 1989. Changing generic strategies: Likehood, direction and performance implications. Strategic Management Journal, 10(5): 413-430.

Zbaracki, M.J., 1998. The retoric and reality of total quality management. Administrative Science Quarterly, 43(3): 602-636.

Zhao, X., A.C.L. Yeung and T.S. Lee, 2004. Quality management and organizational context in selected service industries of China. Journal of Operations Management, 22(6): 575.

Ziggers, G.W. and J.H. Trienekens, 1999. Quality assurance in food and agribusiness supply chains: Developing succesful partnerships. International Journal of Production Economics, 60-61: 271-279.

Appendices

Appendix 1. Validation and reliability assessment

In this appendix the different measurement characteristics of reflective and formative constructs are discussed in Section A1.1. Section A1.2 and A1.3 discusses the validation produces for respectively the formative and the reflective constructs.

A1.1 Measurement characteristics of reflective and formative constructs

Constructs can be distinguished in two types of constructs: *reflective* or *formative* constructs (see Figure A1.1). *Reflective* constructs represent latent variables that cannot be measured directly, but are computed from one or more items. Items are affected by the latent variable. This rationale is represented by the following formula $Y_i = \lambda_{ij}{}^* \eta_j + \varepsilon_i$ in which Y_i is the *i*th item of the latent variable η_j. λ_j is the standardised loading coefficient representing the expected effect of η_j on Y_i and ε_i is the error term of item Y_i (Bollen and Lennox, 1991). As can be seen from the formula the latent variable determines its items, therefore, reflective constructs are also called *effect* constructs.

The nature of *formative* or *causal* constructs is opposite to those of reflective constructs. For formative constructs items can be viewed as causing the variable rather than the items being caused by the variable. Formative constructs are composed of items which directly represent the operational definition and are regarded as explanatory combinations of items. In the case of a single item measure the latent variable η is equal to the empirical measure of the item X (Scholten, 2006). If the formative construct consists of several items, it is specified by the following formula $\eta_j = \gamma_{j1}X_1 + \gamma_{j2}X_2 + \ldots \gamma_{jn}X_n + \xi_j$ in which, γ_{j1} reflects the contribution of X_i to the latent variable η_j and ξ_j is the measurement error (Bollen and Lennox, 1991). In Figure A1.1 formative and reflective constructs are shown.

From Figure A1.1 and the equations which represent formative and reflective constructs. The differences become clear that they have characteristics on which they differ from each other (Fornell *et al.*, 1991):
1. The items in the reflective constructs are interchangeable, which means that leaving out one item does not change the contents of the latent construct. For formative constructs leaving out one item is *omitting a part of the construct* (Bollen and Lennox, 1991). For example, in this study, the construct external pressure is formative, leaving out the driver media attention could undermine the content of the construct.
2. Correlations between the items of formative constructs are not explained by the measurement model and are exogenously determined which makes assessing validity and reliability a problem, because items do not necessarily have to correlate as illustrated in Figure A1.1.

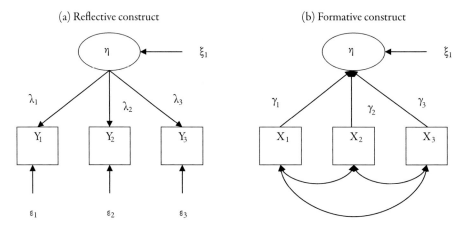

Figure A1.1. Reflective constructs and formative constructs (Bollen and Lennox, 1991).

3. Meaningful items of a formative construct can even be inversely related; there is no reason why the signs and magnitudes of the items should display a specific sign or magnitude. Internal consistency between the items of the formative construct is therefore not important.
4. Formative items have no error terms, but the error in the total construct is represented by the disturbance term ξ which is uncorrelated with the items X.

The different characteristics of the formative and reflective constructs mean that the techniques for validity and reliability of reflective constructs (for example, factor analysis and internal consistency) cannot be used for the validation of formative constructs.

A1.2 Assessing the validity of formative constructs

The validation procedure for the formative construct is different compared to the validation of the reflective constructs, due to the different characteristics of both constructs. Because formative constructs are based on items that represent separate dimensions of a construct items do not necessarily correlate with each other. Therefore, statistical methods for assessing the validity such as convergent validity and discriminant validity cannot be applied (see Section A1.3).

The validation procedures for formative constructs are depicted in Figure A1.2 and include content validity, nomological validity, and item multicollinearity as discussed below.

Content validity

Content validity is concerned to which degree the *domain* of the construct is captured by its measurements, in fact it is concerned with what the construct is measuring (Churchill,

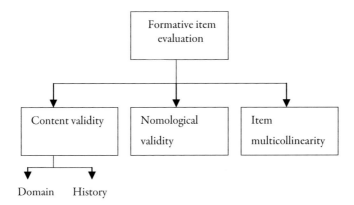

Figure A1.2. Procedure to assess validity and reliability of formative constructs. Adapted form Claro (2003).

1999). The key task in order to achieve content validity is that during the design of the survey attention is paid to ascertaining the domain of the construct. The next step is to gather items that represent the variable as defined. Churchill (1999) recommends that the collection of items must be so large that after refinement of the measurements (based on statistical procedures) there are still enough items to adequately sample the domain of study. A common method within social sciences and used in the present study is to look at the *history* of the scales. If a measurement scale performed well in previous studies, this supports the scale validity. *Domain* validity of the construct is also improved by asking respondents during the pre-test whether or not they think that no important items for the constructs have been omitted, or the other way around non relevant items are included in the survey. The comments of the respondents in the pre-test were used to fine-tune the content validity of the items. A formative construct is directly represented by its items, therefore, the content validity of the domain of the formative construct is very important, because omitting an item omits a part of the construct (Bollen and Lennox, 1991). As a result, the number of items has a strong influence on the breath of the formative construct.

Nomological validity

Nomological validity or criterion validity is concerned with the 'behaviour' of the construct, or in other words, how well it is related to other theoretically related constructs (Churchill, 1999). Nomological validity of reflective constructs is assessed by testing the pre-specified hypotheses in the research model between the construct of interest and other constructs. When satisfactory support is found for the hypotheses, nomological validity of the constructs is achieved (Steenkamp and Van Trijp, 1991).

Item multicollinearity

Item multicollinearity was introduced by Diamantopoulos and Winklhofer (2001) which refers to high correlation between variables of the indicators. The stability of the indicator coefficients (γs see Figure A1.1b) depends on the sample size and strength of the indicators' inter-correlations. Too much multicollinearity between two variables makes it difficult to separate the direct influence of the individual Xs on the variable η. Item multicollinearity was tested by computing Pearson correlations between the items of a construct. According to Hair *et al.* (1998) and Field (2003) Pearson correlations[77] between the items higher than 0.80 are indicators of multicollinearity.

A1.3 Assessing the validity and reliability of reflective constructs

For assessing the validity of reflective constructs many procedures exist as content validity, nomological validity, convergent validity and discriminant validity. Reliability can be assessed by computing the Cronbach's α, composite reliability and variance extracted. Figure A1.3 depicts the procedure to assess validity and reliability. The methods to determine the indicators for the validity and reliability indicators are shown in ellipses.

Content and nomological validity

Like the validation procedure of the formative constructs, the validation procedure of the reflective constructs starts with the *content validity*. The approach for the reflective construct does not differ from that of the formative constructs. Also the approach for testing *nomological validity* of reflective is assessed in the same way as for formative constructs.

Convergent validity

Convergent validity measures to what extent measures correlate positively with other measures of the same construct (Churchill, 1999). For the assessment of convergent validity of reflective constructs item-total correlation, explorative factor analysis (EFA) and confirmatory factor analysis (CFA) were used. Item-total correlation is the correlation of one item with the average

---

[77] The Pearson correlation presents the magnitude and direction of the association between two variables in a data set (Malhotra *et al.* 1999). It is an index used to determine whether a linear or straight-line relationship exists between two variables. The correlation coefficient is a number between -1 and +1, which remains the same regardless of the underlying units of measurements. Calculation of the coefficient considers the mean and the standard deviation of the two variables in the sample (Churchill, 1999). The magnitude of the coefficient is the strength of the correlation. The closer the coefficient is to either -1 or +1 the stronger the correlations. If the correlation coefficient is 0 or very close to 0, there is no association between the two variables. The direction of the correlation specifies how the two variables are related. If the correlation is positive, the two variables have a positive relationship (i.e. as one increases, the other also increases), whereas if the correlation is negative, the two variables have an inverse relationships (i.e. as one increases, the other decreases). The importance of significant correlation coefficients are based on the t-values.

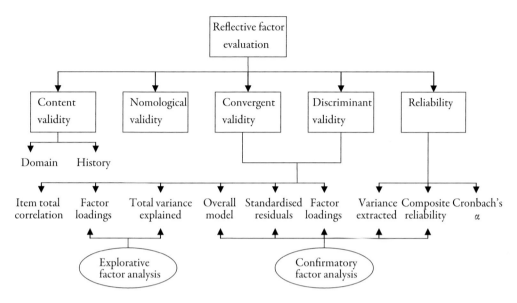

Figure A1.3. Procedures to assess validity and reliability of reflective constructs. Adapted from Claro (2003).

of all other items of a certain construct. The threshold level for item-total correlation is 0.50, because items with lower values do not share a substantial part of the variance with the other items constituting a construct (Steenkamp and Van Trijp, 1991). For assessment of the factor loadings of the items in the EFA and CFA threshold levels of 0.60 were used. To further increase convergent validity, the explained variances were assessed, which should be preferably greater than 60%. Finally the variance extracted (higher than 0.50) from the CFA was used as a last measure for convergent validity.

Discriminant validity

Discriminant validity assesses the extent to which a construct and its indicators differ from other constructs and their indicators. If the correlation between two constructs is very high they might be measuring the same phenomenon instead of different phenomena. Discriminant validity was assessed by three methods. *Firstly,* discriminant validity of a construct is established when the Cronbach's α is larger than the interscale correlations (Ghisellla *et al.*, 1981; Kaynak, 2003). *Secondly*, if the percentage of variance extracted by the indicators of a construct is consistently greater than the average squared inter construct correlations of the construct, discriminant validity of the construct with respect to all other constructs is established (Ahire and Dreyfus, 2000). *Thirdly*, more conservatively, discriminant validity is achieved if each individual correlation of a construct with another construct is lower than its Cronbach's α (Fornell and Larcker, 1981).

Reliability

The *reliability* of the reflective construct refers to the extent that a set of indicators is consistent with its measures. For the reliability of the construct, three measures were used.

Firstly, Cronbach's α is calculated, the most widely used measure of scale reliability (Peterson, 1994). The key assumption behind Cronbach's α is that all items that belong to the domain of a certain construct have common variance. The total variance of a set of items is divided in variances which resemble the true variance in the latent variable and in error variance. The total variation of the set of items that resemble the true variation in the latent variable is equal to Cronbach's α. That means that if all items are drawn form the same construct, they should show high correlations among each other. With the Cronbach's α a summary measure of the inter-correlations that exist among a set of items is provided. Cronbach's α is calculated as (Churchill, 1999):

$$\alpha = \left(\frac{k}{k-1} \right) \left(1 - \frac{\sum_{i=1}^{k} \sigma_i^2}{\sigma_t^2} \right) \qquad \text{(Churchill, 1999)}$$

In which:
k = number of items in the scale
σ_i^2 = variance of scores on item i across subjects
σ_t^2 = variance of total scores across subjects where the total score for each respondent represents the sum of the individual item scores (variance of the scale).

The scores of the Cronbach's α range from 0 (indicating that the items perform very badly in capturing consistency) to 1 (indicating that the items perform very well in capturing consistency). Reported recommended reliability levels differ, but generally accepted levels are 0.60 for explorative studies, and 0.70 for confirmatory studies.

Secondly, the composite or construct reliability of the latent constructs, developed by Werts *et al.* (1974), is calculated on basis of the standardised loadings and standardised errors provided by the output of the CFA. The composite reliability can be calculated by using the following formula:

$$Composite\ reliability = \frac{\left(\sum_{i=1}^{k} \lambda_{ij} \right)^2}{\left(\sum_{i=1}^{k} \lambda_{ij} \right)^2 + \sum_{i=1}^{k} \varepsilon_{ij}} \qquad \text{(Werts et al., 1974)}$$

In which
k = number of items of the scale
λ_{ij} = standardised loading coefficient i of the path from the observed to the latent variable j
ε_{ij} = error term of coefficient i to the latent variable j

---

Box A1.1. Confirmatory factor analysis.

Confirmatory Factor Analysis (CFA) stems from econometrics and merged with the principles of measurement from psychology and sociology. It is a widely accepted method in managerial and academic research (Hair *et al.*, 1998). In CFA the measurement model was estimated and assessed in the statistical program Lisrel 8.72. This model estimates the relations between the observed items and the latent constructs, but does not estimate any structural paths (relationships between the latent constructs). Figure A1.1a shows a measurement model of a latent variable. After separate examination of all the measurement models of the constructs, a CFA on the total measurement model including all the constructs of both supplier and the buyer model was carried out. For CFA theoretical justification is a key word, because the model has to be specified completely by the researcher and its goal is to confirm (theoretically) pre-specified relationships. This is the opposite perspective of the EFA, which identifies common factors and explains the relationships to the observed data (Lattin *et al.*, 2003). The derived structure of the EFA is data driven and can be different from the structure that could be expected from theory.

During the analyses of the CFA in this study the covariance matrix was used and was preferred above the correlation matrix. The use of the covariance matrix is strongly recommended in almost all instances and especially for structural models, the next step after the CFA. Structural models are not always scale free which means that models that will fit on the correlation matrix may not fit on the covariance matrix. Hypothesis testing available in structural equation modelling (SEM) assumes a covariance matrix (Kelloway, 1998; Lattin *et al.*, 2003). Moreover, when multi-group analyses are performed (see Section 5.3.4) one should use the covariance matrix, otherwise no valid comparisons between models can be made; correlation matrices remove important information about the scale of measurement of individual variables from the data (Baumgartner and Homburg, 1996; Hair *et al.*, 1998).

During the analysis Maximum Likelihood Estimation (MLE) was used, an estimation method that obtains good results with relatively small sample sizes. The MLE technique produces the best estimations if the observed variables have a normal distribution. Therefore, in the present study the kurtosis and the skewness of the variables were assessed and should not exceed |1|.

CFA offers the possibility to modify the model to give a better representation of the empirical data. However, adjustment of the model should never be done at random, but only on theoretical justifications, because adjustment lowers the meaning and the substantial conclusions that can be drawn from the model (Gerbing and Anderson, 1984). It is recommended that the standardised λ of each indicator should be greater than 0.60 and its accompanying t-value greater than 1.96. To asses how well the specified model accounts for the data, overall goodness of fit indices should be used, see also Box 5.2 (Anderson and Gerbing, 1988; Steenkamp and Van Trijp, 1991; Hair *et al.*, 1998; Steenkamp and Baumgartner, 1998).

For the composite reliability, a recommended threshold level is 0.70. Despite Cronbach's α, this measure does not assume equivalency among the measures with its assumption that all indicators are equally weighted. It is less sensitive to the number of items of the construct. In CFA, the composite reliability of an item is defined as the direct relationship between the latent variable and the item. The larger the relationship, the higher the reliability of an item Y_i (see Figure A1.1).

Thirdly the variance extracted, developed by Fornell and Larcker (1981) can also be calculated based on the standardised loadings and standardised errors. The formula for the variance extracted is shown below:

$$Variance\ extracted = \frac{\sum \lambda_{ij}^2}{\sum \lambda_{ij}^2 + \sum_{i=1}^{P} \varepsilon_{ij}} \qquad \text{(Fornell and Larcker, 1981)}$$

In which
k = number of items of the scale
λ_{ij} = standardised loading coefficient i of the path from the observed to the latent variable j
ε_{ij} = error term of coefficient i to the latent variable j

Although in explorative factor analysis it is recommended that 60% of the variance is explained by the factor solution, a recommended level in confirmatory analysis is 50%. Fornell and Larcker (1981) suggested that this measure can also be interpreted as a more conservative measure of reliability for the latent variable.

In Table A1.1 the threshold levels of the various evaluation criteria for the validity and reliability of the reflective constructs are summarised.

Table A1.1. Summary of the statistical evaluation criteria for reflective constructs.

Evaluation criteria	Threshold
Validation of construct	
Inter-item total correlation	≥ 0.50
Explorative factor analysis	
Explained variance	≥ 0.60
Factor loadings	≥ 0.60
Confirmatory factor analysis	
Standardised loadings (λ)	≥ 0.60
t-value of the standardised loadings	≥ 1.96
Reliability of the constructs	
Cronbach's α	≥ 0.60
Composite reliability	≥ 0.70
Composite validity (variance extracted)	≥ 0.50

Appendix 2. Analyses of the sub-groups

Although it could be concluded from Chapter 7 that to a large extent the data could be aggregated across groups of firms, in this appendix the data is investigated in detail for several sub-groups of firms. The total sample was divided into a number of sub-groups in order to find further evidence of the applicability of the research model[78] and to compare the groups of firms with each other. Moreover, a number of control variables were added to the models. Three different kinds of sub-groups were created and within these sub-groups the following firms were compared:
1. Primary producers from different chains
2. Traders/processors from different chains
3. Primary producers and traders/processors

Several reasons plead for using multiple regression instead of Structural Equation Modelling (SEM) in this part of the study. *Firstly*, because some control variables are specific for certain sub-groups it is not possible to study them in a multi group analysis. Some fit indices of the sub-models were somewhat below the threshold levels, likely due to the small group sizes. *Secondly*, some control variables are binary variables, which make them less suitable for SEM which is very sensitive for deviations of multi-normality (Hair *et al.*, 1998). *Thirdly,* adding the testing of the effect of control variables would result in many non-significant paths being added to the model resulting in a decrease of fit. All these reasons taken together would suggest that that SEM would not give the best representation of the results in this case.

For each firm in a sub-group the supplier and the buyer model were estimated on the basis of a series of five regression equations. Dependent variables were the integration of quality management with suppliers by the focal firm, commitment of the suppliers, enforcement by the focal firm, satisfaction of the focal firm with suppliers and revenue growth of the focal firm for the supplier model. For the buyer model the dependent variables were integration of quality management with the focal firm by the buyers, commitment of the focal firm, enforcement of the buyers, buyer satisfaction with the focal firm and revenue growth of the focal firm for the buyer model.

The multiple regression analyses were carried out in two stages. In the *first* stage only control variables were used as predictors for the independent variables which explain a part of the dependent variable that does not necessarily have to do with independent variables defined in the research model. In the *second* stage both control variables and research variables together were included in the multiple regression analyses as predictors for the dependent variables.

---

[78] It is allowed to use this model across the groups, because in the previous section metric invariance, factor (co) variance invariance and path invariance were demonstrated. This means that across the groups the factors are comparable. Moreover, a double check was carried out by conducting an explorative factor analyses for each group separately to test if the same factors emerged, which was indeed the case. Furthermore, the Cronbach α's for the constructs in the sub groups were satisfying.

Multiple regression reports for both stages the relevant statistics such as R^2, R^2 adjusted and their significance (F-value) and the standardised coefficients with their significance (t-values). In the second stage the R^2, R^2 adjusted and the F change are compared with the first stage which informs whether or not adding research variables in the multiple regression analyses would make sense to explain the variances of the dependent variables. Indicators for multicollinearity such as the Variance Inflation Factor (VIF, should be lower than 10) and condition indices (lower than 30) were also checked, but did not indicate problems for multicollinearity during the regression analyses. In this chapter only the outcomes of the second stage are discussed, because the second stage reflects the model of interest.

Between the sub-groups also the magnitudes of the scores on the constructs were compared to investigate if sub-groups differ from each other. To compare the scores on the constructs, a univariate analysis of variance (ANOVA)[79] was carried out. In the case that only two groups were compared with each other a t-test was used. After the ANOVA post hoc tests were carried out to compare all groups with each other in order to find which groups differ from each other. In this study the Bonferroni, Hochberg's GT2 and the Games Howell tests were used. All these tests take specific circumstances of samples into account. Bonferroni is suitable when the number of comparisons is small and is conservative, Hochberg's GT2 takes different sample sizes into account and Games Howell has been designed for comparing groups with different variances (Field, 2003). However, during the analyses no differences in the outcomes of the various post hoc tests were found.

A2.1 Primary producers

The first sub-groups consist of the primary producers from the poultry meat chain, the fruit and vegetable chain and the flower and potted plant chain. Included control variables in the regression analyses for the primary producers are size, presence of a chain leader, presence of a quality manager, extra quality management systems above the standard quality system and young owner or successor present. If a control variable turned out to be significant in the regression analysis in a certain group it is mentioned in the box besides the figure see for example Figure A2.1 together with the direction of the effect. The scores of the groups on the constructs of both the supplier and buyer model are depicted in Table A2.1 In the last column the means of the groups are compared (The F in the last column means average).

- It is remarkable that the scores for the primary producers in the flower and potted plant chain are significantly lower on many quality management related constructs (environmental pressure, quality strategy, TSIs in both models, integration of quality management in both models and buyer satisfaction in the buyer model) compared to the other two chains.

---

[79] Analysis of variance (ANOVA) is used to test if several means (more than two) are equal and is an extension of the two-sample t test (Field, 2003). Post hoc tests are run after the ANOVA have been conducted in order to find out which means of sub-groups differ from each other. For example, the ANOVA can indicate that the the means of the of sub-groups A, B and C are not equal. With the post-hoc test it is possible to find out whether A or B differ from each other, or A and C or B and C.

Table A2.1. Means and standard deviations of the constructs of both the supplier and the buyer models for primary producers.

	Poultry meat Mean (St. dev.)	Fruit and vegetables Mean (St. dev.)	Flowers and potted plants Mean (St. dev.)	Comparison of means[1]
Both models				
External pressure on focal firm	57.2 (20.6)	57.3 (20.2)	47.8 (19.4)	$F_{FP} < F_{FV}, F_P$
Information exchange by ICT	3.4 (1.7)	4.3 (1.6)	3.7 (1.6)	$F_{FV} > F_P, F_{FP}$
Quality strategy of focal firm	5.2 (1.1)	5.3 (1.1)	4.7 (1.2)	$F_{FP} < F_P, F_{FV}$
Revenue growth of focal firm	4.5 (1.1)	4.0 (1.3)	4.2 (1.3)	$F_P > F_{FV}$
Supplier model				
TSIs by suppliers	4.3 (1.6)	3.9 (1.6)	2.9 (1.6)	$F_{FP} < F_P, F_{FV}$
Integration of quality management with suppliers	4.7 (1.2)	4.2 (1.3)	3.7 (1.3)	$F_{FP} < F_P, F_{FV}$
Commitment by suppliers	5.5 (1.2)	4.9 (1.5)	4.9 (1.5)	$F_P > F_{FP}, F_{FV}$
Enforcement by focal firm	4.3 (1.5)	4.0 (1.6)	3.8 (1.7)	$F_P > F_{FP}$
Satisfaction of focal firm about suppliers	5.4 (1.2)	5.2 (1.0)	4.8 (1.1)	$F_P > F_{FP}$
Buyer model				
TSIs by focal firm	4.4 (1.4)	5.2 (1.1)	3.8 (1.7)	$F_{FV} > F_P, F_{FP}$; $F_P > F_{FP}$
Integration of quality management with buyers	4.8 (1.2)	5.0 (1.2)	3.6 (1.4)	$F_{FP} < F_P, F_{FV}$
Commitment of focal firm	5.6 (0.9)	5.9 (0.9)	5.7 (1.1)	-
Enforcement by buyers	4.8 (1.4)	5.2 (1.2)	4.5 (1.6)	$F_{FV} > F_{FP}$
Satisfaction of buyers about focal firm	6.0 (1.0)	6.0 (0.9)	5.6 (1.1)	$F_{FP} < F_P, F_{FV}$

[1] $p \leq 0.05$.
F_P = Average score of poultry.
F_{FV} = Average score of fruit and vegetables.
F_{FP} = Average score of flowers and potted plants.

- Moreover, primary producers from the poultry meat chain have significantly higher scores for commitment in the supplier model.
- The primary producers in the fruit and vegetable chain have higher scores on information exchange by ICT than other chains, but achieve a lower revenue growth than poultry

farmers. Poultry farmers also have higher score on enforcement and satisfaction than growers of flowers and potted plants in the supplier model.
- Growers of flowers and potted plants have lower scores on enforcement in the buyer model than growers of fruit and vegetables.

In Figure A2.1 and Figure A2.2 the outcomes of the regression analyses for respectively the supplier and buyer model for the primary producers are displayed.
- Except for the flower and potted plant chain in the buyer model, external pressure has a significant positive impact on the integration of quality management.
- For all primary producers TSIs have a significant effect on integration of quality management in both models.
- Information exchange by ICT has no effect on the integration of quality management with buyers for all primary producers and also no effect on the integration of quality management with suppliers in the fruit and vegetable chain.
- For primary producers, the quality strategy of the focal firm behaves to a large extent like it does in the total models, because there is only an effect on integration with the suppliers, except in the flower and potted plants chain.
- Commitment is significantly related to integration of quality management in both models except in the supplier model of the poultry meat chain and in the buyer model of the flower and potted plant chain.
- It is remarkable that many research variables have no effect on the integration of quality management with the focal firm by the buyer in the flower and potted plant sector, but for this group the presence of a quality manager in the own firm is important for the integration of quality management in the supplier and the buyer model.

The integration of quality management has a positive significant effect on the commitment in the chain for all primary producers in both models, except for the flower and potted plant chain in the buyer model. These results hardly differ from the results in the general models. Also the effect of the integration of quality management on enforcement is significant in both models except in the buyer model in the fruit and vegetable sector. For this group the regression equation was not significant, which could be the result of the low reliability of enforcement in this group. For enforcement some control variables have a significant effect, but there is no specific control variable systematically related to enforcement in all groups.

Commitment has a positive significant effect on buyer satisfaction in both models for all the primary producers. Although in the total supplier model, integration of quality management has a positive effect on buyer satisfaction, it is not present in the sub-group of primary producers. Less statistical power due to much lower numbers of firms included in the regression analysis is a likely explanation. It is remarkable that firms with a young owner or a successor achieve a higher buyer satisfaction in the supplier model of the flower and potted plant chain.

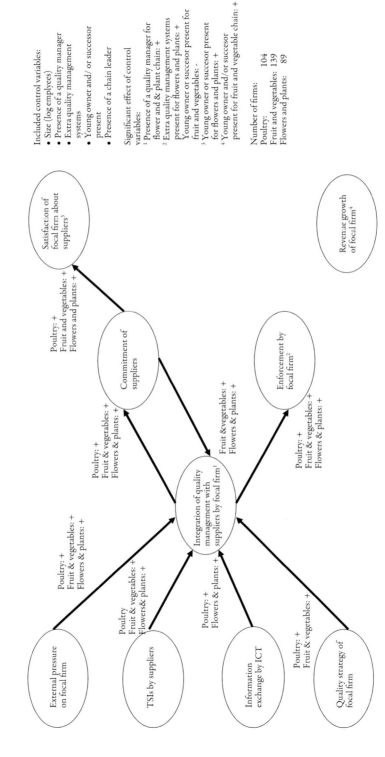

Included control variables:
- Size (log emplyees)
- Presence of a quality manager
- Extra quality management systems
  - Young owner and/ or successor present
  - Presence of a chain leader

Significant effect of control variables:
[1] Presence of a quality manager for flower and & plant chain: +
[2] Extra quality management systems present for flowers and plants: +
Young owner or succesor present for fruit and vegetables: -
[3] Young owner or succesor present for flowers and plants: +
[4] Young owner and/ or succesor present for fruit and vegetable chain: +

Number of firms:
Poultry: 104
Fruit and vegetables: 139
Flowers and plants: 89

Figure A2.1. The relationships between the most important suppliers and the focal firm for the primary producers. Only paths that are significant on p ≤ 0.05 in at least one group are displayed.

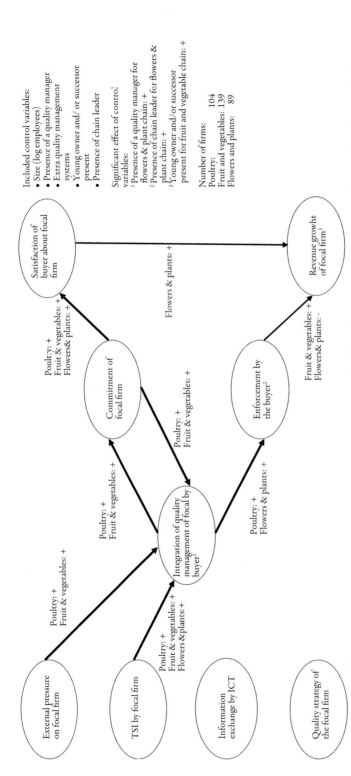

Figure A2.2. The relationships between the most important buyers and the focal firm for the primary producers. Only paths that are significant on p ≤ 0.05 in at least one group are displayed.

As in the total model, revenue growth of the focal firm is difficult to explain with the research variables included in this study, because many of them they have no effect on the revenue growth in both models for the primary producers. There is a remarkable positive effect of enforcement on revenue growth in the buyer model of the primary producers in the fruit and vegetable chain. Regarding the low reliability of this construct for this group, the results should be interpreted with care. Furthermore, there is a negative effect of enforcement on revenue growth in the buyer model for the flower and potted plant chain. In the fruit and vegetable chain, firms with younger owners or with a successor achieve higher revenue growth. Other control variables have no effect at all on the revenue growth of primary producers.

From these partial analyses it can be concluded that many significant relationships in both the supplier and buyer model were significant in the sub-models for the primary producers too. Less statistical power due to a lower number of firms in the sub-models is likely the most important reason why some relationships are not significant in the sub-models. Moreover, control variables did not play an important role in both the supplier and buyer models for primary producers.

A2.2 Traders and/or processors

The second sub-groups consist of the traders and the processors from the fruit and vegetable chain and from the flower and potted plant chain. The traders and/or processors of the poultry meat chain were included in the ANOVA, but were not included in the regression analyses, because the number of firms (34) was too small to carry out reliable regression analyses. Included control variables in the regression analyses for the traders and/or processors are size, presence of a chain leader, presence of a quality manager, extra quality management systems and number of suppliers and buyers for respectively the supplier and buyer model.

The scores of these firms on the constructs in both the buyer and supplier model are shown in Table A2.2.
- As for the primary producers in the flower and potted plant chain, also traders from this chain have significant lower external pressure from the business environment. The integration of quality management with the buyers for this group is also lower compared to the other two groups.
- Traders from the flower and potted plant chain also have a significant lower level of integration of quality management with the suppliers than the fruit and vegetable traders.
- Also of interest is that traders from the flower and potted plant chain achieve higher revenue growth and are more aimed at information exchange by ICT in their business relationships than traders from other chains.
- Moreover, in the flower and potted plant chain the score of TSIs is significantly lower than in the fruit and vegetable chain in the supplier model and significantly lower than in the poultry meat chain in the buyer model.

Table A2.2. Means and standard deviations of the constructs of both the supplier and the buyer models for traders and/or processors.

	Poultry meat Mean (St. dev.)	Fruit and vegetables Mean (St. dev.)	Flowers and potted plants Mean (St. dev.)	Comparison of means[1]
Both models				
External pressure on focal firm	59.4 (19.1)	54.4 (18.0)	44.9 (21.9)	$F_{FP} < F_{FV}, F_P$
Information exchange by ICT	3.2 (1.9)	3.3 (1.9)	4.2 (1.9)	$F_{FP} > F_P, F_{FV}$
Quality strategy of focal firm	5.6 (1.0)	5.0 (1.5)	5.1 (1.4)	-
Revenue growth of focal firm	3.8 (1.6)	3.8 (1.2)	4.5 (1.1)	$F_{FP} > F_{FV}, F_P$
Supplier model				
TSIs by suppliers	4.0 (1.5)	4.2 (1.5)	3.4 (1.7)	$F_{FP} < F_{FV}$
Integration of quality management with suppliers	4.5 (1.2)	4.6 (1.2)	4.1 (1.2)	$F_{FV} > F_{FP}$
Commitment by suppliers	6.0 (0.9)	5.6 (1.1)	5.8 (1.0)	-
Enforcement by focal firm	4.6 (1.7)	4.8 (1.6)	4.9 (1.5)	-
Satisfaction of focal firm about suppliers	5.7 (0.9)	5.3 (1.1)	5.3 (1.1)	-
Buyer model				
TSIs by focal firm	5.0 (1.2)	4.4 (1.6)	3.9 (1.8)	$F_{FP} < F_P$
Integration of quality management with buyers	4.5 (1.3)	4.4 (1.1)	3.8 (1.4)	$F_{FP} < F_P, F_{FV}$
Commitment of focal firm	6.1 (0.9)	6.0 (0.9)	6.1 (1.1)	-
Enforcement by buyers	5.1 (1.7)	4.8 (1.5)	4.4 (1.5)	-
Satisfaction of buyers about focal firm	6.0 (0.9)	5.9 (0.9)	5.9 (0.9)	-

[1] $p \leq 0.05$

F_P = Average of poultry

F_{FV} = Average of fruit and vegetables

F_{FP} = Average of flowers and potted plants

- Remarkable is also that for both models there are no significant differences on the scores between traders from the poultry meat chain and the fruit and vegetable chain.

In Figure A2.3 and Figure A2.4 outcomes of regression analyses on integration of quality management for respectively the supplier and buyer model are shown.

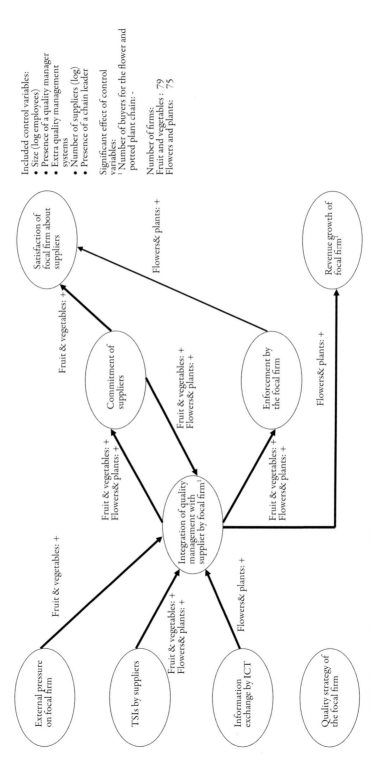

Figure A2.3. The relationships between the most important suppliers and the focal firm for the traders and or processors. Only paths that are significant on $p \leq 0.05$ in at least one group are displayed.

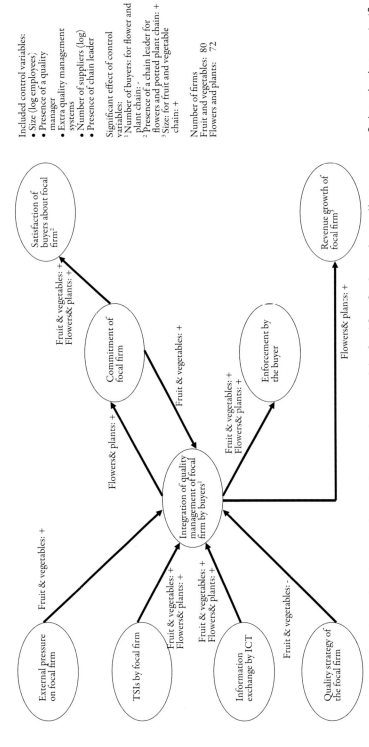

Figure A2.4. The relationships between the most important buyers and the focal firm for the traders and/or processors. Only paths that are significant on $p \leq 0.05$ in at least one group are displayed.

- For traders in the flower and potted plant chain external pressure has no significant effect on integration of quality management in both models, strengthening the vision that external pressure is less important in this chain.
- It is remarkable that information exchange by ICT has no effect on integration of quality management with suppliers in the fruit and vegetable chain. It is also remarkable that quality strategy of the focal firm has no effect on the integration of quality management with the suppliers for traders.
- Very striking is the significant negative effect of the quality strategy of the focal firm on the integration with the buyer in the fruit and vegetable chain, because although not significant in other chains, the direction of the relationships is positive. In the supplier model there were weakly significant relationships ($p \leq 0.10$) between quality strategy and the integration of quality management for the fruit and vegetable chain and the flower and potted plant chain.
- Another deviation from the total model is that commitment has no effect on the integration of quality management in the buyer model for the flower and potted plant chain.
- Except the negative effect of the number of buyers on the integration of quality management in the buyer model for flowers and potted plant chain control variables have no effect on the integration of quality management.

Furthermore, integration of quality management has a positive effect on the commitment in both models, except in the buyer model of the fruit and vegetable traders. Control variables have no effect on commitment in both models. Results for the effect of integration of quality management on enforcement are in line with the results of overall models.

The integration of quality management with suppliers and buyers has no significant effect for traders on buyer satisfaction. It is remarkable that for traders of flowers and potted plants commitment has no significant effect on buyer satisfaction in the supplier model, but enforcement has a positive effect, which is quite opposite to the fruit and vegetable chain. Control variables hardly play any role for buyer satisfaction in both models, only the presence of a chain for traders and/or processors in the flower and potted plant chain. A very interesting result is that the integration of quality management has a significant effect on revenue growth in both the supplier and the buyer model for traders in the flower and potted plant chain. This effect is not present for traders in the fruit and vegetable chain. Commitment and enforcement have no significant effect on revenue growth in both models and for both kinds of traders. Bigger firms in the fruit and vegetable chain realise higher revenue growth compared to smaller firms in the buyer model, because the control variable size has a positive effect on revenue growth.

The models for the traders represent to a large extent the overall models. Most control variables do not have important impacts on the dependent variables during the regression analyses.

A2.3 Primary producers versus traders and/or processors

In the third analysis, the primary producers are compared with the traders and/or processors. The scores of these two groups on the constructs of the research model are displayed in Table A2.3. The number of differences on the scores of the constructs is small, however, there are some differences:

- Traders and/or processors achieve significantly higher scores on commitment and enforcement in the supplier model than the primary producers.
- It is remarkable that traders and/or processors are closer to the end-buyer, but receive a lower score on the integration of quality management and TSIs in the buyer model.

Table A2.3. Means and standard deviations of the construct of both the supplier and the buyer models for primary producers and traders/processors.

	Primary producers Mean (St.dev.)	Traders/ Processors Mean (St.dev)	Comparison of means[1]
Both models			
External pressure on focal firm	54.6 (20.5)	51.5 (20.5)	-
Information exchange by ICT	3.8 (1.7)	3.6 (1.9)	-
Quality strategy of focal firm	5.1 (1.2)	5.1 (1.4)	-
Revenue growth of focal firm	4.1 (1.3)	4.1 (1.4)	-
Supplier model			
TSIs by suppliers	3.8 (1.7)	3.7 (1.7)	-
Integration of quality management with suppliers	4.2 (1.3)	4.4 (1.2)	-
Commitment by suppliers	5.1 (1.5)	5.8 (1.0)	$F_{PP} < F_{TP}$
Enforcement by focal firm	4.0 (1.6)	4.8 (1.6)	$F_{PP} < F_{TP}$
Satisfaction of focal firm about suppliers	5.2 (1.1)	5.3 (1.1)	-
Buyer model			
TSIs by focal firm	4.6 (1.5)	4.3 (1.7)	$F_{PP} > F_{TP}$
Integration of quality management of focal firm	4.5 (1.4)	4.2 (1.3)	$F_{PP} > F_{TP}$
Commitment of focal firm	5.9 (1.0)	6.0 (1.0)	-
Enforcement by buyers	4.9 (1.4)	4.7 (1.5)	-
Satisfaction of buyers about focal firm	5.9 (1.0)	5.9 (0.9)	-

[1] $p \leq 0.05$.

F_{PP} = Average of primary producers.

F_{TP} = Average of traders/processors.

The outcomes for the regression analyses for the supplier and buyer model for primary producers and traders and/or processors are depicted in respectively Figure A2.5 and Figure A2.6. Included control variables in the regression analyses are size, presence of a chain leader, presence of quality manager, extra quality management systems and chain.

The outcomes of this regression analyses are very close to the outcomes of the overall models, although there are some small differences. The only difference is that information exchange by ICT has no effect on the integration of quality management in the buyer model of primary producers. Some control variables also had significant effects. For primary producers there was a significant positive effect of the presence of a quality manager on the integration of quality management with the suppliers. In the buyer model the flower and potted chain had a negative impact on the integration of quality management.

The results for commitment show expected outcomes compared to the overall model, in all cases commitment was positively related to integration of quality management. In the supplier model, commitment was also positively related to the poultry meat chain for both primary producers and traders and/or processors. The flower and potted plant chain has a positive effect on commitment of traders both in the supplier and buyer model. The results for enforcement are for both models in line with the overall models. For the primary producers there is positive relationship between the presence of a chain leader and enforcement in the buyer model.

Only in the supplier model of traders, integration of quality management is positively related to buyer satisfaction. In both models for both primary producers and traders and/or processors commitment and enforcement are not positively significant related to buyer satisfaction which is in line with the overall models. The control variable presence of a chain leader has a positive effect on buyer satisfaction in the buyer model of traders and/or processors.

Integration of quality management, flower chain and size seem to be important for the revenue growth of traders, because in both the supplier and the buyer models these predictor variables are significant. For the primary producers the presence of a quality manager and the poultry meat chain are important.

Also for these sub-groups a lot of support for both the supplier and buyer model is found. These sub-groups in particular are very alike the total models, because many relationships are significant that are also significant in the total supplier and buyer model. This strengthens the vision that relationships that are not significant in sub-groups, but are significant in the overall models, are mainly results of less statistical power, because these groups in this analysis are considerably greater than the previous sub-groups.

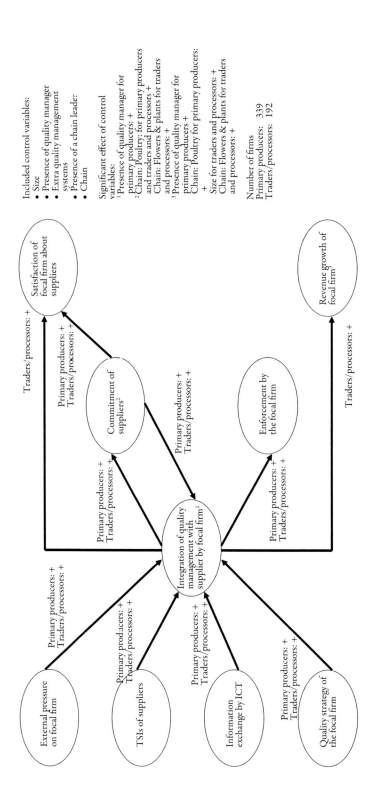

Figure A2.5. The relationships between the most important suppliers and the focal firm for all primary producers and all traders and/or processors. Only paths that are significant on p ≤ 0.05 in at least one group are displayed.

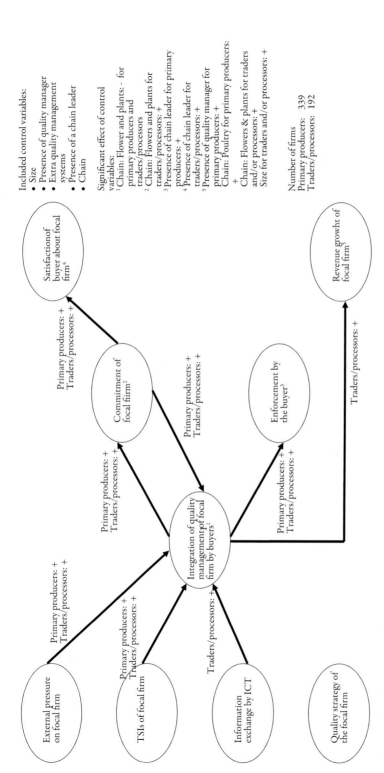

Figure A2.6. The relationships between the most important buyers and the focal firm for all primary producers and all traders and/or processors. Only paths that are significant on p ≤ 0.05 in at least one group are displayed.

A2.4 Perceptions of the relationships

In this section two sequential linkages in each chain are investigated. For each linkage a supplier and buyer model has been developed. This implies that it is possible to compare the perception about the relationships that exists between suppliers and buyers, see Figure A2.7.

In order to investigate this, the scores of the buyer and suppliers about the relationships were compared. The results are shown in Table A2.4.

Respondents agree on the level of integration of quality management, commitment and enforcement, but disagree on the level of TSIs and satisfaction. Suppliers think that they have done more TSIs than their buyer thinks they have done. Furthermore, suppliers think that their buyers are more satisfied then they really are. An explanation is that people perceive themselves more positively than others do, thus having an over-optimistic view of themselves [80]. However,

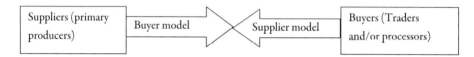

Figure A2.7. Perceptions about the relationships, the scores in the supplier model of traders deal with the same relationships as the scores in the buyer model of primary producers.

Table A2.4. The perceptions of the relationships between suppliers (primary producers) and buyers (traders and/or processors).

	Suppliers Mean (st. dev.)	Buyers Mean (st. dev.)
TSIs	4.6 (1.1)[a]	3.8 (1.7)[a]
Integration of quality management	4.5 (1.4)	4.4 (1.2)
Commitment	5.9 (1.0)	5.8 (1.0)
Enforcement	4.9 (1.4)	4.8 (1.5)
Buyer satisfaction	5.9 (1.0)[a]	5.3 (1.1)[a]

[a] Significant differences between primary producers and traders and/or processors about the relationship ($p \leq .05$).

---

[80] A more detailed analysis showed that in the poultry chain respondents agree on all scores, the same was true for the flower and plant chain, except for the integration of quality management. However in the fruit and vegetable chain respondents disagree on all scores.

the results should be interpreted with care because there is no one to one relationship between the primary producers and traders and/or processors in the study sample. In practice this means that it is not possible to make conclusions about individual relationships, only about general patterns in relationships.

Also the differences in the perceptions about what the suppliers and buyers think have an impact on the integration of quality management could be compared. The outcomes are depicted in Figure A2.8.

For most relationships the perceptions are quite close to each other, for example for the effect of external pressure, TSIs and quality strategy of the focal firm, the opinions are equal. For information exchange by ICT, buyers indicate that it has a significant positive effect on the integration of quality management with suppliers. A possible explanation could be that buyers have a larger size and have much more suppliers and integrated information exchange may support the supplier management of the buyers. Quality strategy has no effect according to suppliers as opposed to the buyers. For a firm it is much easier to impose its own management strategy on its suppliers than on its buyers.

For the effect of integration of quality management on commitment different perceptions between suppliers and buyers do not exist. The same holds for enforcement. Regarding the results for buyer satisfaction, it is remarkable that buyers think that a higher buyer satisfaction could be directly achieved by more integration of quality management. In the questionnaire the questions has been formulated in such a way that the buyers initiate the integration of quality management. Therefore, if it does not contribute to the satisfaction of buyers, they would not be likely to put any efforts into making the integration working. For the factors that have an effect on revenue growth, the only difference is that buyers again think that integration of quality management directly contributes to more revenue growth. The same explanation that buyers will not put any effort into integration activities which will not result in a higher revenue growth is a likely explanation too.

Taking all dependent variables into account, it can be stated that suppliers and buyers are share to a large extent their visions about which indicators have an effect on the dependent variables of their relationship.

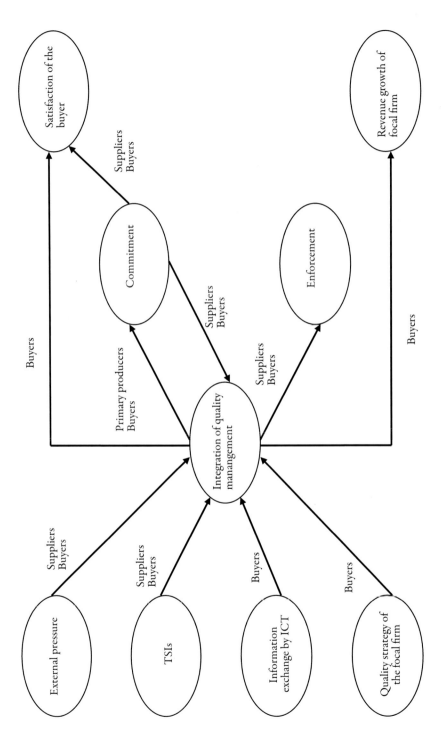

Figure A2.8. Perception about the relationships between suppliers (primary producers) and buyers (traders and/or processors).

Appendix 3. The drivers and their levels included in ACA

The general environment

Media attention
1. The chances are high that our products will be linked to negative media attention.
2. The chances are low that our products will be linked to negative media attention.

Supra-national legislative demands
1. Export companies increasingly have to comply with more stringent quality legislation from outside the EU.
2. Import companies increasingly have to comply with more stringent EU quality legislation.

National legislative demands
1. Government interacts more frequently with industry organisations on quality issues.
2. Government interacts less frequently with industry organisations on quality issues.

Changing consumer demands

For the poultry meat and fruit and vegetable chain:
1. Consumers increasingly demand more convenient foods.
2. Consumers increasingly demand more exotic foods.
3. Consumers increasingly demand more healthy foods.

For the flower and potted plant chain:
1. Consumers increasingly demand more convenient flowers.
2. Consumers increasingly demand more exotic flowers.
3. Consumers increasingly demand more flowers with a long vase life.

Societal demands on corporate social responsibility
1. Our chain partners increasingly concentrate on corporate social responsibility (e.g. animal welfare, labour conditions) programmes.
2. Our chain partners concentrate less on corporate social responsibility (e.g. animal welfare, labour conditions) programmes.

Willingness to pay for a quality label
1. Consumers are increasingly willing to pay extra for a quality label.
2. Consumers are NOT willing to pay extra for a quality label.

Globalisation of import
1. Import comes increasingly from developing countries.
2. Import comes increasingly from developed countries.

Globalisation of export
1. Export goes increasingly to developing countries.
2. Export goes increasingly to developed countries.

Different quality regulations/systems
1. Quality regulations are increasingly harmonised among countries.
2. Quality regulations are increasingly differentiated among countries.

The task environment

Chain-wide innovation in quality management systems
1. Innovations for quality improvements are increasingly developed through vertical co-operation in the chain.
2. Innovations for quality improvements are increasingly developed through horizontal co-operation in the chain.
3. Innovations for quality improvements are increasingly developed by individual companies in the chain.

Information exchange by ICT
1. The use of ICT in transfer of quality data (e.g. product specifications) increases quickly.
2. The use of ICT in transfer of quality data (e.g. product specifications) increases slowly or does not increase.

Increasing power dependency in the chain
1. One or a few companies in the chain increasingly enforce quality demands (including certification systems) within the chain.
2. There is no company in the chain which enforces quality demands (including certification systems) within the chain.

Appendix 4. The questionnaire survey

This appendix reports the items used in the survey, their scales and their Cronbach α's.

Both models

Quality strategy of the focal firm (7-point Likert scale, '1=not applicable at all', '4 = neutral', '7=totally applicable'). Cronbach α = 0.81.

The management of our firm:

QS1:	supports initiatives (from inside and outside the firm) to improve quality management.
QS2:	regards quality management as necessary for the firm's performance.
QS3:	regards quality management as one of the top priorities in evaluating the firm's performance.
QS4:	thinks that quality is more important than price *(item dropped)*.

External pressure on focal firm (4 different sub groups, 7-point Likert scale, '1=not applicable at all', '4 = neutral', '7=totally applicable').

Media attention:

During the last three years media attention with regard to quality meant for our firm:

Media1:	an important driver affecting our quality activities.
Media2:	an effective means of competing in our market.
Media3:	a threat for our revenues.

Legislative demands:

During the last three years legislation with regard to quality meant for our firm:

Legis1:	an important driver affecting our quality activities.
Legis2:	an effective means of competing in our market.
Legis3:	a threat for our revenues.

Changing consumer demands:

During the last three years changing consumer demands with regard to quality meant for our firm:

CCD1:	an important driver affecting our quality activities.
CCD2:	an effective means of competing in our market.
CCD3:	a threat for our revenues.

Appendices

Societal demands for corporate social responsibility:

During the last three years corporate social responsibility meant for our firm:
CSR1:   an important driver affecting our quality activities.
CSR2:   an effective means of competing in our market.
CSR3:   a threat for our revenues.

Integrated information exchange (7-point Likert scale, '1=not applicable at all', '4 = neutral', '7=totally applicable').

ICT1:   Our firm is principally aimed at the use of ICT for information transfer with other chain partners.

Revenue growth of focal firm (7-point Likert scale, '1 = much slower', '4 = neutral', '7 = much faster').

Grow1:   Compared to our main competitors our revenues grow:

Specific supplier model constructs

Transaction specific investment by supplier (7-point Likert scale, '1=not applicable at all', '4 = neutral', '7=totally applicable'). Cronbach = 0.91.

In order to comply with our quality requirements, our most important suppliers have largely invested:
TsiS1:   in production means.
TsiS2:   in working procedures.
TsiS3:   in administration and information structure.
TsiS4:   in adapting their quality management to ours.

Integration of quality management with suppliers by focal firm

Monitoring (7-point Likert scale, '1=not applicable at all', '4 = neutral', '7=totally applicable'). Cronbach α = 0.91.

Our firm requires from our most important suppliers:
MonS1:   frequent (e.g. every 3 months) information about the outcomes of specific quality tests and – inspections.
MonS2:   easy access to the procedures for quality assurance (e.g. ICT systems).
MonS3:   active participation in the monitoring system for the quality assurance of our firm.

Alignment (7-point Likert scale, '1=not applicable at all', '4 = neutral', '7=totally applicable').
Cronbach α = 0.88.

In the relationship with our most important suppliers our firm:
AlignS1:   strives for a close collaboration in order to align the quality processes.
AlignS2:   makes special appointments to achieve a better quality performance.
AlignS3:   communicates the quality requirements clearly and precisely.
AlignS4:   strives for long-term relationships *(item dropped)*.

Improvement (7-point Likert scale, '1=not applicable at all', '4 = neutral', '7=totally applicable').
Cronbach α = 0.85.

In the relationship with our most important suppliers our firm:
ImproS1:   periodically provides feedback about quality performance.
ImproS2:   provides every kind of information that the supplier might need in order to comply with our quality requirements.
ImproS3:   passes the quality requirements of our buyers frequently and effectively.
ImproS4:   measures buyers' satisfaction and relates this to the quality performance of our most important suppliers.

Commitment of suppliers (7-point Likert scale, '1=not applicable at all', '4 = neutral', '7=totally applicable'). Cronbach α = 0.90.

Our most important suppliers:
ComS1:   know our quality requirements.
ComS2:   regard our quality requirements as reasonable.
ComS3:   feel responsible to comply with our quality requirements.
ComS4:   comply with little effort (in terms of time and money) to our quality requirements *(item dropped)*.

Enforcement by focal firm (7-point Likert scale, '1=not applicable at all', '4 = neutral', '7=totally applicable'). Cronbach α = 0.76.

Our most important suppliers:
EnfS1:   are more frequently controlled if they do not comply with our quality requirements.
EnfS2:   regard the sanctions as severe if the they do not comply with our quality requirements.
EnfS3:   think that non compliance with our quality requirements damages their image *(item dropped)*.

Satisfaction of focal firm about suppliers (7-point Likert scale, '1=not applicable at all', '4 = neutral', '7=totally applicable'), Cronbach α =0.85.

SatS1:	Our firm is satisfied about the quality of the products that our most important suppliers deliver.
SatS2:	Our firm trusts the quality of the processes of our most important suppliers.
SatS3:	Our most important suppliers manage to comply rapidly with changing quality requirements of our firm.

Specific buyer model constructs

Transaction specific investments by focal firm (7-point Likert scale, '1=not applicable at all', '4 = neutral', '7=totally applicable'). Cronbach α = 0.87.

In order to comply with the quality requirements of our most important buyers, our firm has largely invested:

TsiC1:	in production means.
Tsic2:	in working procedures.
TsiC3:	in administration and information structure.
TsiC4:	in adapting our quality management to theirs.

Integration of quality management with focal firm by buyer

Monitoring (7-point Likert scale, '1=not applicable at all', '4 = neutral', '7=totally applicable'). Cronbach α = 0.90.

Our most important buyers require from our firm

MonC1:	frequent (e.g. every 3 months) information about the outcomes of specific quality tests and – inspections.
MonC2:	easy access to the procedures for quality assurance (e.g. ICT systems).
MonC3:	active participation in the monitoring system for the quality assurance of their firm.

Alignment (7-point Likert scale, '1=not applicable at all', '4 = neutral', '7=totally applicable'). Cronbach α = 0.88.

Our most important buyers:

AlignC1:	strive for a close collaboration with our firm in order to align the quality processes.
AlignC2:	make special appointments with our firm to achieve a better quality performance.

AlignC3:   communicate the quality requirements to our firm clearly and precisely.
AlignC4:   strive for long-term relationships with our firm *(item dropped)*.

Improvement (7-point Likert scale, '1=not applicable at all', '4 = neutral', '7=totally applicable'). Cronbach α = 0.87.

Our most important buyers:
ImproC1:   periodically provide feedback to our firm about quality performance.
ImproC2:   provides every kind of information that our firm might need in order to comply with their quality requirements.
ImproC3:   pass the quality requirements of their buyers frequently and effectively to our firm.
ImproC4:   measure buyers' satisfaction and relate this to the quality performance of our firm.

Commitment of focal firm (7-point Likert scale, '1=not applicable at all', '4 = neutral', '7=totally applicable'). Cronbach α = 0.83.

Our firm:
ComC1:   knows the quality requirements of our most important buyers.
ComC2:   regards the quality requirements of our most important buyers as reasonable.
ComC3:   feels responsible to comply with the quality requirements of our most important buyers.
ComC4:   complies with little effort (in terms of time and money) to the quality requirements of our most important buyers *(item dropped)*.

Enforcement by buyers (7-point Likert scale, '1=not applicable at all', '4 = neutral', '7=totally applicable'). Cronbach α = 0.65.

Our firm:
EnfC1:   is more frequently controlled if it does not comply with the quality requirements of our most important buyers.
EnfC2:   regards sanctions as severe if it does not comply with the quality requirements of our most important buyers.
EnfC3:   thinks that non compliance with the quality requirements of our most important buyers damages our image *(item dropped)*.

Satisfaction of buyers about focal firm (7-point Likert scale, '1=not applicable at all', '4 = neutral', '7=totally applicable'). Cronbach $\alpha = 0.84$.

SatC1:	Our most important buyer is satisfied about the quality of the products which the firm delivers.
SatC2:	Our most important buyers trust the quality of the processes of our firm.
SatC3:	Our firm manages to comply rapidly with changing quality requirements of our most important buyers.

Control variables

Quality manager

Does your firm employ one or more quality managers? (yes/no) If yes how many?

Size

How many persons (own personnel and hired personnel, full time equivalents) does your firm employ at the moment?

What is the turn-over of your firm per year (millions of Euros)?

Chain leader

Is there a firm present in your chain that is able to enforce it's quality requirements on your chain? (yes/no) If yes who?

Number of quality management systems:

For which quality systems is your firm certified?

Number of suppliers and buyers (only traders and processors)

What is the number of suppliers and buyers of your firm?

_____suppliers

_____buyers

Most important buyer (only traders and processors)

To which kind of firm does your most important buyer belong to? (Retailers and other categories)

Age (only primary producers)

What is your age (years)

Successor (only primary producers)

Does your firm have a successor? (yes, no, unknown)

Appendix 5. Interview protocol for the in-depth interviews

Respondent profile

- Name of the respondent
- Function of the respondent
- Time spent within the firm
- Date and place of interview

'Best practice' quality management systems

1. What is new / different / characteristic to the quality management system of firms that perform above average on quality management within the firms themselves and in relation with their most important suppliers and buyers compared to other firms, for example, their competitors?
2. According to you, what are the most important critical success factors and bottlenecks for obtaining a 'best practice' quality management within a firm and with their most important suppliers and buyers?
3. How did firms that perform above average on quality management deal with external pressures for better quality such as *media attention, legislative demands, changing consumer demands* and *societal demands for corporate social responsibility* within their firm and in relation with their most important suppliers and buyers? Which pressures did emerge? Do you have examples about the way firms deals with these requirements?
4. Did quality management of firms that perform above average lead to other advantages (or possible disadvantages) than improved quality only for these firms or in relation to their most important suppliers and buyers?
5. Did firms that perform above average on quality management encounter problems with the government or other parties during the execution of quality management within their firm or in relation with their most important suppliers and buyers?

Self regulation

1. Is it possible, and if yes, in which way should the government shape self regulation for quality assurance in your industry?
2. What are according to you the most important critical success factors for obtaining optimal level of self regulation for quality assurance in your industry?
3. What are according to you the most important advantages and disadvantages of self regulation of quality assurance in your industry?
4. Which roles should the government, industry organisation, Product Boards and / or certifying organisation play for establishing self regulation of quality assurance? Which roles do they play now? Which roles should they have to play in the future? How important are these organisations? What could be improved according you?

5. The government is not necessary for quality assurance; the market can do it much better. A right proposition? Or is the government necessary in some cases? If yes, in what cases and why?

Summary

Introduction

During the last decade, concerns about quality and safety in agri-food supply chains have been raised. Several sector-wide crises, such as the BSE and the dioxin crises, classical swine fever and foot-and-mouth disease and Aviaire Influenza have fuelled these concerns. These concerns may not only be limited to safety and quality issues, but also important ethical concerns are raised, for example, concerning preventive slaughtering of animals. Due to all this attention, consumers have become more critical regarding the food products they buy. The EU and the national governments have also reacted on the above mentioned crises by setting up regulations for quality and safety of agri-food products. Furthermore, retailers have introduced quality management standards, such as BRC and Eurep-GAP in which they impose quality requirements on their suppliers.

In order to comply with the quality requirements of the government and the retailers, closely coupled agri-food supply chains have emerged. The essence of such closely integrated chains is to create collaboration and commitment in which partners share information, work together to solve problems, jointly plan for the future and make their success interdependent. Due to higher transparency firms are also better able to enforce quality requirements. However, concerns have been raised about the (administrative) burdens being placed on firms, because at the moment firms have to comply with many different private and public quality regulations. In order to reduce the compliance burdens, the Dutch Ministry of Agriculture, Nature and Food Quality wants to implement a new inspection policy, called 'control-on-control', whereby the private sector is assigned more responsibility for compliance with statutory regulations. The government operates at a greater distance, but retains the ultimate responsibility.

Research questions

Building and maintaining good businesses relationships between partners are daunting tasks. Therefore, understanding the factors that determine successful collaboration and integration on the one hand and self regulating of quality management systems in agri-food supply chains on the other hand is very important. Furthermore, it is important to find the right balance between the two important dimensions of self regulation, commitment and enforcement, which enables 'control-on-control'. To address the challenges described above, three research questions are formulated:
1. Which (internal and external) factors have an impact on the integration of quality management systems in agri-food supply chains?
2. How do integrated quality management systems affect self regulation and performance in agri-food supply chains?
3. What is the best way to create self regulated quality management systems in agri-food supply chains?

Although research interest in supply chain management is clearly growing, only a few studies have been directed to quality management practices in a supply chain perspective. The present study deals with this research gap by investigating the integration and self regulation of quality management in agri-food supply chains.

Research model

In this study a number of theories have been used. The Supply Chain Management (SCM) and Total Quality Management (TQM) theories are used to define the most important elements of quality management in a supply chain perspective. Due to intensive collaboration in the chain, for example, on quality management, higher performance for the individual firms in the chain is expected. Literature on buyer-supplier relationships frequently states that increased performance is likely to be best achieved by means of committed suppliers and buyers. For measuring performance of a firm buyer satisfaction and revenue growth of the firm were used. In addition, SCM emphasises the importance of information exchange by means of ICT which can be regarded as a catalyst for successful integration of supply chain processes. Transaction Cost Theory underlines the impact of transaction specific investments (TSIs) as needed for the integration of quality management. Due to strong collaboration in chains supported by these investments, opportunistic behaviour of the chain partners is to a large extent prevented. Furthermore, external drivers (*media attention, legislative demands, changing consumer demands* and *societal demands for corporate social responsibility*) put pressure on firms to integrate their quality management. From a Contingency Theory perspective a firm that faces more external pressures will be more inclined to integrate its quality management systems with its suppliers and buyers whereby the importance of the own quality strategy for the success of this integration is emphasised. As was already argued, the integration of quality management with suppliers and buyers is expected to have a positive impact on self regulation (commitment and enforcement), because firms get common goals and transparency in the chain, which enables them better to control each other. These thoughts are summarised in Figure S1.

An important feature of the study is that it collects data from both the supplier and the buyer side of the firm and includes two successive stages in each chain (primary producers and traders and/or processors). This approach ensures the proper implementation of the SCM approach. Until now, most studies were limited to data collection in the firm or solely about the suppliers or buyers. In this study, the research model was applied to the supplier and buyer side of the focal firm and for each model hypotheses were formulated (Table S1).

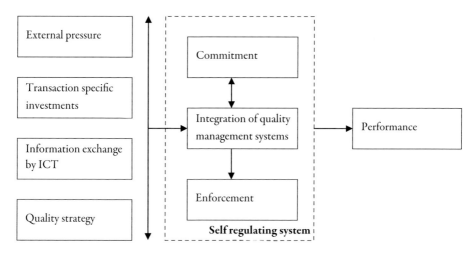

Figure S1. Theoretical model.

Study domain

The research model was tested in the poultry meat chain, the fruit and vegetable chain and the flower and potted plant chain, because:
1. These chains are valid representations of the agri-food sector, they are characterised by a large diversity of marketing channels and products.
2. All these chains are of great interest for the Dutch economy, especially with regard to export
3. All the three chains pay a lot of attention to quality management.

Methodology

The study makes use of 'mixed methodology' combining qualitative and quantitative research methods. The study was carried out in three phases. The *first* phase of this study starts with the identification, description and ranking of external pressures acting on agri-food supply chains, with experts from business and research. These interviews were combined with a conjoint analysis, a quantitative method to arrive at a ranking of the importance of the different drivers/ pressures for integration of quality management and self regulation in the three chains. In the *second* phase a survey was conducted among primary producers, processors and/or traders in the three chains. The primary goal of the survey was to test hypotheses of the theoretical model. In the *third* phase, the findings from the quantitative part of the research were verified using in-depth interviews with experts from the three chains. By doing so, for example, the advantage of the generalisibility of the questionnaire can be combined by greater insight in the way of working of firms in their single, natural setting, provided by the in-depth interviews. Besides, on the basis of the results, examples of 'best practices' about quality management and self regulation were formulated for managers and policy makers.

Summary

Table S1. The hypotheses for the supplier and buyer model. S refers to the supplier model, B to the buyer model.

Hypotheses

1   The higher the pressure from the business environment with regard to product and process quality, the higher the level of integration of quality management systems between the firm and its suppliers (H_1S) and its buyers (H_1B).

2   The higher the level of transaction specific investments of the suppliers cq the firm, the higher the level of integration of quality management systems between the firm and its suppliers (H_2S) or its buyers (H_2B).

3   The higher the level of information exchange in the chain by means of ICT, the higher the level of integration of quality management systems between the firm and its suppliers (H_3S) or its buyers (H_3B).

4   The more the firm pursues its quality strategy, the higher the level of integration of quality management systems between the firm and its suppliers (H_4S) and its buyers (H_4B).

5   The higher the level of commitment of the suppliers to the quality requirements of the buyers, the higher the level of integration of quality management systems between the suppliers and buyers (H_{5a}). And vice versa: The higher the level of integration of quality management systems between the suppliers and buyers), the higher the level of commitment of the suppliers to quality requirements of the buyers (H_{5b}).

6   The higher the level of integration of quality management systems between the firm and its suppliers or buyers the higher the possibilities for enforcement of the quality requirements by the firm (H_6S) cq the buyers (H_6B).

7   The higher the level of integration of quality management systems between the firm and its suppliers (H_7S) or its buyers (H_7B), the higher the buyer satisfaction will be and ultimately the financial performance of the focal firm.

Results and conclusions

In the *first* phase of the study, 47 interviews with experts from business and research were held. The results of the conjoint analysis show a clear ranking of the drivers across different chains. Drivers such as *media attention, changing consumer demands* and *societal demands for corporate social responsibility* in particular have an important impact on the on-going integration of quality management systems of firms, while pressures such as the globalisation of trade were regarded as less important.

In total 585 firms reacted to the survey. Table S2 shows the distribution of the respondents across the chains involved. Based on the analysis of the survey, it turned out that the general research model was highly generalisable for the different kinds of firms involved in the study.

Table S2. Number of firms per chain.

Firm	Number of firms			
	Poultry meat	**Fruit and vegetables**	**Flowers and potted plants**	**Total**
Primary producers	116	151	102	369
Traders/processors	34	98	84	216
Total	150	249	186	585

Therefore, it could be regarded as a robust model for studying quality management and self regulation in agri-food supply chains. However, the measured level of quality management in the flower and potted chain was significantly lower compared to the poultry meat en the fruit en vegetable chain. An explanation is that food safety does not play a role in the flower and potted plant chain.

Regarding the *first* research question, this study has shown that if firms perceive stronger external pressures, their quality management systems will be more integrated with suppliers and buyers (see Figure S2). Many of the pressures are not aimed at specific firms, but often influence all firms in a supply chain. However, incorrect actions of only one firm in the supply chain may result in increasing external pressures on all firms in the chain. By integrating quality management systems in agri-food supply chains, managers try to prevent this. Interestingly, *legislative demands* have hardly any impact on the integration of quality management

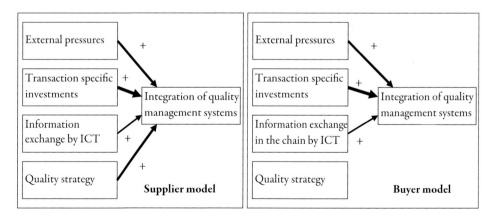

Figure S2. Factors influencing integration of quality management with the supplier (left) and buyer (right); no line means no significant relationship.

systems with buyers and suppliers, while *media attention, societal demands for corporate social responsibility* and *changing consumer demands* have a great impact.

TSIs and integrated ICT systems (for example 'tracking and tracing systems') also contribute to the successful integration of quality management systems. Integration and collaboration on quality lowers the risks for opportunistic behaviour. Firms send an important signal to other parties in the chain that the relationship is highly valued by TSIs and integration of quality management systems. Interestingly, the quality strategy of the focal firm has an impact on the integration of quality management with the suppliers, but not with the buyers. The most likely explanation is that firms are able to impose their quality requirements upstream, but not downstream in the chain. When selecting suppliers, the firm is able to let its interest for quality management play an important role, whereas this is much more difficult in the choice of its buyers.

Regarding the *second* research question, the study showed empirical evidence that integrated quality management systems are strongly positively related to self regulation, see Figure S3. Due to the integration of quality management systems a platform is established for open communication about specifications and chain process improvements which results in a mutual understanding and commitment for each other's quality requirements. Moreover, the exchanges of outcomes of quality test and - inspections results in more possibilities for enforcement of quality requirements.

This study also shows that integration of quality management leads to higher performance. Firms that have integrated their quality management systems with their suppliers and buyers achieve higher levels of performance (both for buyer satisfaction and revenue growth). This effect is achieved by commitment of the parties in the chain and not by means of enforcement. A policy that is focused too much on enforcement and sanctions has no effect in the supplier model and works even detrimentally in the buyer model. Enforcement has the potential to result in conflicts with suppliers, especially if sanctions are imposed that are perceived to be unjust or unreasonable. However, although a large majority of firms will comply with quality requirements as well as possible and too strong enforcement is de-motivating for them, a certain level of enforcement is needed for firms that will behave opportunistically.

Recommendations for managers and policy makers

In order to answer the *third* research question based on 14 in-depth interviews important recommendations are formulated for managers and policy makers for the creation of self regulated quality management systems in agri-food supply chains. The results from the first and the second phase of the study are also included in these recommendations. The first recommendations are intended to establish self regulated quality management systems in firms:
1. Due to the use of quality management systems, managers have access to an abundant source of information about the quality performance of their own firm and their suppliers, which

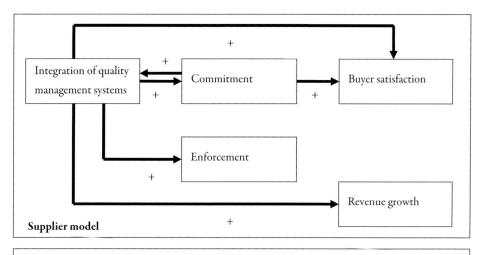

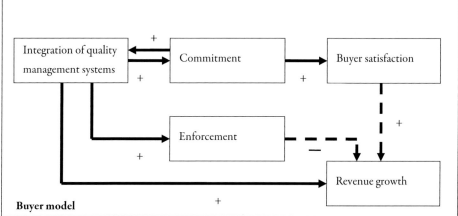

Figure S3. Impact of integration of quality management on self regulation and performance of firms in agri-food supply chains (dashed arrow means weakly significant relationship).

is often only used to verify compliance. Analysing this often underused or even unused data can result in new insights in performance and may lead to (great) process improvements.

2. Managers should take care that quality management is 'alive' in the firm. If quality management is integrated well with all business functions, such as the human resource management policy of the firm, and it serves as a means supportive to the design of superior firm and chain processes, it will not be regarded as merely a bureaucratic burden.

The study has also derived a number of important implications for policy makers. These recommendations mainly focus on facilitating and improving self regulation in agri-food supply chains.

Summary

1. To minimise the chance of fraud, inspection procedures should be clearly described and supervised by an independent Council of Accreditation. This prevents the commercial relationship between audited firms and certifiers influencing certifiers in their evaluation. 'Control-on-control' may then even be more effective in preventing fraud. Certifiers know firms and develop relationships with them which are aimed at the improvement of quality management systems.

2. As a result of 'control-on-control' the effectiveness of governmental inspections will increase, because the government is 'fishing where the fishes swim' (bad performing firms on quality are inspected more frequently). The government should stress the higher effectiveness of the 'control-on-control' approach on an EU-level, because it might falsely suggest a high number of non-compliances in The Netherlands. Fair comparisons are extremely important for Dutch agri-food supply chains, regarding the large export interests.

3. Not *legislative demands*, but other pressures such as *'media attention'* and *'societal demands for corporate social responsibility'* have an impact on the integration of quality management. Therefore, policy makers should not focus exclusively on prescriptive legislative, but on innovative approaches to increase the integration of quality management systems to motivate firms to deliver safe and high quality foods, for example, publishing the names of poorly performing firms.

About the author

Wijnand van Plaggenhoef was born in Nijkerkerveen on the 3rd of January 1979. In 1997, he received his VWO certificate at the Van Lodensteincollege in Amersfoort. In the same year he entered Wageningen University and Research Centre to study Agricultural Engineering. During his study, he became fascinated with supply chain management in agri-food supply chains. This inspired him to choose a specialisation in management and logistics. After an internship at RIKILT, the Institute of Food Safety in The Netherlands, he received his MSc. in 2003 and started his PhD research. He has presented his work at various international conferences and contributed to book chapters. Besides the PhD research, he was involved in a number of international research projects such as the EU concerted action Global Food Network and the Interreg IIIC project Promotion the Stable-to-Table APproach (PromSTAP). Furthermore, from 2003 to 2007 he has been the secretary of 'The Journal on Chain and Network Science', a peer-reviewed scientific research journal, which aims to promote theory and practice in the field of innovation in business chains and networks. The journal is intended to serve the needs of academics and management scientists, practising managers, students, research personnel and consultants. Since October 2007 Wijnand van Plaggenhoef joined Significant in Barneveld (an independent spin-off of PriceWaterhouseCoopers) as a consultant.

Printed in the United States
by Baker & Taylor Publisher Services